PRAISE FOR
Newton and the Counterfeiter

"Levenson's account of this world of criminality, collusion, and denunciation is meticulously researched and highly readable . . . the tale of Newton the economist is one worth telling."
— *New Scientist*

"Levenson demonstrates a surpassing felicity in his brisk treatment of this late-17th-century true-crime adventure . . . Swift, agile treatment of a little known but highly entertaining episode in a legendary life." — *Kirkus Reviews*

"*Newton and the Counterfeiter* packs a wonderful punch in its thoroughly surprising revelation of that other Isaac Newton, and in its vivid re-creation of 17th-century London and its fascinating criminal haunts." — *Providence Journal*

"A pacy and absorbing history thriller . . . Levenson's book is a gripping tale of unrelenting revenge and obsession." — *Financial Times*

"Levenson reveals the remarkable and true tale of the only criminal investigator who was far, far brainier than even Sherlock Holmes: Sir Isaac Newton during his tenure as Warden of the Royal Mint. What a fascinating saga!"
— Walter Isaacson, author of *Einstein: His Life and Universe* and *Benjamin Franklin: An American Life*

"A wonderful read that reveals a whole new side to a giant of science. Through a page-turning narrative, we witness Isaac Newton's genius grappling with the darker sides of human nature, an all too human journey reflecting his deepest beliefs about the cosmic order. This is a gripping story that enriches our sense of the man who forever changed our view of the universe."
— Brian Greene, author of *The Fabric of the Cosmos*

"As the great Newton recedes from us in time, he comes increasingly into focus as a man rather than a myth—thanks in no small measure to this learned and lively new study from the estimable Thomas Levenson."—Timothy Ferris, author of *Seeing in the Dark*

"*Newton and the Counterfeiter* is both a fascinating read and a meticulously researched historical document: a combination difficult to achieve and rarely seen ... Recommended for anyone who wants to know the real story behind this astonishing but largely overlooked chapter of scientific history."
—Neal Stephenson, author of *Cryptonomicon* and *Anathem*

"I absolutely loved *Newton and the Counterfeiter*. Deft, witty, and exhaustively researched, Levenson's tale illuminates a near-forgotten chapter of Newton's extraordinary life—the cat-and-mouse game that pitted him against a criminal mastermind—and manages not only to add to our knowledge of the great mathematician but to make a page-turner out of it. This book rocks."
—Junot Díaz, author of *The Brief Wondrous Life of Oscar Wao*

"*Newton and the Counterfeiter* is a rollicking account of the fascinating underbelly of seventeenth-century London—and reveals an aspect of Newton I'd scarcely known of before, yet which shaped the world we know. A tour de force."
—David Bodanis, author of *E=mc2*

Newton and the Counterfeiter

BOOKS BY THOMAS LEVENSON

*Ice Time: Climate, Science,
and Life on Earth*

*Measure for Measure: A Musical
History of Science*

Einstein in Berlin

*Newton and the Counterfeiter:
The Unknown Detective Career of
the World's Greatest Scientist*

Newton
and the
Counterfeiter

❦

The Unknown Detective Career of
the World's Greatest Scientist

Thomas Levenson

⚓

Mariner Books
Houghton Mifflin Harcourt
BOSTON   NEW YORK

For Henry
who added years to the writing
and joy to the years
(as your grandfather once wrote
in a similar context)

&

for Katha, always

First Mariner Books edition 2010

Copyright © 2009 by Thomas Levenson

www.hmhbooks.com

Library of Congress Cataloging-in-Publication Data
Levenson, Thomas.
Newton and the counterfeiter : the unknown detective career
of the world's greatest scientist / Thomas Levenson.
p. cm.
Includes bibliographical references and index.
ISBN 978-0-15-101278-7
1. Newton, Isaac, Sir, 1642–1727. 2. Chaloner, William. 3. Counterfeits
and counterfeiting—England—History—17th century. I. Title.
Q143.N495L48 2009
530.092—dc22 [B]   2008053511
ISBN 978-0-547-33604-6 (pbk.)

Book design by Brian Moore

Printed in the United States of America

DOM 10 9 8 7 6 5 4 3 2 1

CONTENTS

~◦◦◦~

"Let Newton Be"

IN EARLY FEBRUARY 1699, a middle-ranking government official found himself a quiet corner of the Dogg pub. He was dressed appropriately. After almost three years on the job, he knew better than to dress for the Royal Society when he wished to pass unremarked in Holborn or Westminster.

The pub was, he hoped, a place where two men could speak discreetly. Big as London was, it could still be a very small town. Men employed in a given trade—legitimate or otherwise—tended to know one another.

The man he awaited came in. His companions would have had to hang back, keeping an eye on their charge from a distance. The newcomer knew the rules—as he should—given his current address: Newgate Jail.

The jailbird sat and started to speak.

There was someone, he said, he had been getting close to, a man who liked to talk. That man was cagey, and smart enough not to trust entirely those with whom he spoke—naturally enough given the nature of his companions, who, like him, were all awaiting trial. But after weeks and months in the cells, staring at the same faces, the monotony of prison life had got to him, and there was not much else to do but talk.

The official listened, increasingly impatient. What had the cellmate said? Did the informer have anything really worth hearing?

No, not quite…perhaps. There was a tool, an engraved plate—you know?

The official knew.

It was hidden, the informer said—of course, for that was what he had been placed in the cell to learn: not just that the plate was hidden, but where.

It was not necessary to remind the jailbird that he lived or died at the official's choice.

The plate is hidden, the informer said, inside a wall or a hollow at one of the houses William Chaloner had last used for a run of counterfeit cash.

Which one?

He didn't know, but Chaloner had boasted that "it was never lookt for in such vacan[t] places."

The detective swallowed his irritation. He already knew that Chaloner was no fool. What he wanted now was something he could get his hands on.

The jailers picked up the hint. It was time to return their charge to Newgate, with orders to do better.

When they were gone, the other man left the pub on his own. He made his way back into the heart of the city, entering the Tower of London through the main west gate.

He turned left and crossed into the precincts of the Royal Mint. There he returned to his usual routine, interrogating another witness, reading over depositions, checking the confessions to be signed.

It was all part of the job, to weave a chain of evidence strong enough to hang William Chaloner—or any counterfeiter whom Isaac Newton, Warden of the Royal Mint, could discover.

Isaac Newton? The founder of modern science; the man recognized by his contemporaries—and ever since—as the greatest natural philosopher the world has ever seen? What had the man who had brought order to the cosmos to do with crime and punishment, the flash world of London's gin houses and tenements, bad money and worse faith?

Isaac Newton's first career, the only one most people remember, lasted thirty-five years. Throughout that period, he was a seemingly permanent fixture at Trinity College, Cambridge—first as a student, next as a fellow, and finally as Lucasian Professor of Mathematics. But in 1696 Newton came to London to take up the post of Warden of the Royal Mint. By law and tradition, the position required him to protect the King's currency, which meant that he was supposed to deter or capture anyone who dared to clip or counterfeit it. In practice, that made him a policeman—or rather, a criminal investigator, interrogator, and prosecutor rolled into one.

A more surprising candidate for the job would be hard to imagine. Newton, in both popular memory and the hagiography of his own time, did not get his hands dirty. He did not so much live as think—and he thought in realms far above those reached by ordinary minds. Alexander Pope captured contemporary sentiment about him in a famous couplet:

> *Nature and Nature's laws lay hid in night:*
> *God said, Let Newton be! and all was light.*

Newton lived, or was imagined to live, beyond the passions and chaos of daily life. It did not take long for his successors to canonize him as a saint in the transforming church of reason. It was no accident that on a 1766 visit to London, Benjamin Franklin commissioned a portrait of himself that shows him sitting at a desk, studying, while a bust of Newton watches over him.

Yet despite having no training or experience or evident interest in the management of men or things, Newton excelled as Warden of the Mint. He tracked, arrested, and prosecuted dozens of coiners and counterfeiters during the four years of his tenure. He knew—or rather, he learned very quickly—how to tangle his opponents in intricately woven webs of evidence, careless conversation, and betrayal. London's underworld had never confronted anyone like him, and most of its members were utterly unprepared to do battle with the most disciplined mind in Europe.

Most, but not all. In William Chaloner Newton found an adversary capable of challenging his own formidable intelligence. Chaloner was no petty criminal. His claimed production of thirty thousand pounds in counterfeit money represented a true fortune—as much as four million pounds in today's currency. He was literate enough to submit pamphlets on finance and the craft of manufacturing coins to Parliament and cunning enough to avoid prosecution for at least six years of a very ambitious criminal career. He was ferocious to a fault, with at least two deaths to his credit, and a profit made from each. Most of all, he was bold. He accused the new Warden of incompetence, even alleged fraud in his management of the Mint. Thus joined, the battle between them raged for more than two years. Before it was over, Newton had made of his pursuit of Chaloner a masterpiece of empirical research. And as he did so he revealed a persona at once less familiar and more coherent, more truly human than the Newton of the hagiographies—a man who not only propelled the transformation in ideas called the scientific revolution but who, along with his contemporaries, lived, thought, and felt them, day in and day out.

That transformation happened both within and to Isaac Newton. To become the man who could run the infamous Chaloner to ground, Newton had to master the habits of mind required for the task. That process, the making of perhaps the most unlikely detective on record, can be dated to the day a young man walked through the gates of a small town in Lincolnshire to further his education.

Part I

Learning to Think

I

"Except God"

June 4, 1661, Cambridge.

The tower of Great St. Mary's catches what daylight remains as a young man passes the town boundaries. He has come about sixty miles, almost certainly on foot (his meticulously kept accounts show no bills paid to livery stables). The journey from rural Lincolnshire to the university has taken him three days. The walls of the colleges shadow Trumpington Street and King's Way, but at this late hour, Trinity College is closed to visitors.

The young man sleeps that night at an inn, and the next morning he pays eight pence for the carriage ride to the college. A few minutes later, he passes beneath the Gothic arch of Trinity's Great Gate and presents himself to college officials for the usual examination. Their scrutiny does not take long. The records of the College of the Holy and Undivided Trinity for June 5, 1661, register that one Isaac Newton has been admitted into its company.

On its face, Newton's entrance to Trinity could not have been more ordinary. He must have seemed to be yet another example of a familiar type, a bright farm youth come to university with the aim of rising in the world. This much is true: now nineteen, Newton was indeed country-bred, but by the time he set foot in Trinity's Great Court it was apparent that he was deeply unsuited for rural life. And he would prove to be a student unlike any the college had ever encountered.

Nothing in his beginnings suggested any such promise. On Christmas Day, 1642, Hannah Newton gave birth to a son, who was so premature that his nurse recalled that at birth he could fit into a quart jug. The family waited a week to christen him with the name of his father, dead for three months.

The infant Isaac was at least reasonably well off. His father had left an adequate landholding, including a farm whose owner enjoyed the grand title of Lord of the Manor of Woolsthorpe. For the time being, however, the inheritance fell to baby Isaac's mother, who was soon able to remarry up. Hannah's second husband, a local clergyman named Barnabas Smith, had a church living, a considerable estate, and admirable energy for a man of sixty-three; he would produce three children with his new wife over the next eight years. There was, it seemed, no place for an inconvenient toddler in such a vigorous marriage. A little more than two years old, Newton was abandoned to the care of his grandmother.

Of necessity, the child Newton learned how to live within his own head. Psychoanalysis at a distance of centuries is a fool's game, but it is a matter of record that, with one possible exception, the adult Newton never permitted himself real emotional dependence on another human being. In the event, his upbringing did not dull his brain. He left his home and village when he was twelve, moving a few miles to the market town of Grantham to begin grammar school. Almost immediately it became obvious that his intelligence was of a different order from that of his classmates. The basic curriculum—Latin and theology—barely troubled him. Contemporaries recalled that when, from time to time, "dull boys were now & then put over him in form," he simply roused himself briefly "& such was his capacity that he could soon doe it & outstrip them when he pleas'd."

In between such interruptions, Newton pleased himself. He drew eagerly, fantastically, covering his rented room with images of "birdes beasts men & ships," figures that included copied portraits of King Charles I and John Donne. He was fascinated by mechanical inventions, and he was good with tools. He built water mills for his own amusement and dolls' furniture for the daughter of his landlord. Time fascinated him: he designed and constructed a water clock, and made sundials so accurate that his family and neighbors came to rely on "Isaac's dials" to measure their days.

Such glimpses of an eager, practical intelligence come from a handful of anecdotes collected just after Newton's death, some seventy years after the event. A closer look can be gained in the notebooks he kept, the first surviving one dating to 1659. In tiny handwriting (paper was precious) Newton recorded his thoughts, questions, and ideas. In that earliest volume he wrote down methods to make inks and mix pigments, including "a colour for dead corpes." He described a technique "to make birds drunk" and how to preserve raw meat ("Immers it in a well stopt vessel under spirits of wine"—with the hopeful postscript "from whose tast perhaps it may be freed by water"). He proposed a perpetual motion machine, along with a dubious remedy for the plague: "Take a good dose of the powder of ripe Ivie berrys. After that the aforesd juice of horse dung." He became a pack rat of knowledge, filling page after page with a catalogue of more than two thousand nouns: "Anguish. Apoplexie. . . . Bedticke. Bodkin. Boghouse. . . . Statesman. Seducer. . . . Stoick. Sceptick."

The notebook contains other lists as well—a phonetic chart of vowel sounds, a table of star positions. Fact upon fact, his own observations, extracts cribbed from other books, his attention swerving from "A remedy for Ague" (it turns on the image of Jesus trembling before the cross) to astronomical observations. The mind emerging on the pages is one that seeks to master *all* the apparent confusion of the world, to bring order where none was then apparent.

At sixteen, though, Newton had no idea how to reconcile

his abilities to his place in life. An exercise notebook from his school days provides a glimpse of real misery. It is a unique document, the purest expression of despair Newton ever committed to paper. He sorrows for "A little fellow; My poore help." He asks: "What imployment is he fit for? What is hee good for?"—and offers no answer. He rails, "No man understands me," and then, at the last, he collapses: "What will become of me. I will make an end. I cannot but weepe. I know not what to do."

Newton wept, but his mother demanded her due. If Isaac had exhausted what his schoolmaster could teach him, then it was time to come home and get back to what should have been his life's work: tending sheep and raising grain.

Let the record show that Isaac Newton made a miserable farmer. He simply refused to play the part. Sent to market, he and a servant would stable their horses at the Saracen's Head in Grantham and then Newton would disappear, making a beeline for the cache of books at his former landlord's house. Or "he would stop by the way between home & Grantham & lye under a hedg studying whilst the man went to town & did the business." On his own land he paid no more attention to his duties. Instead, he "contrived water wheels and dams" and "many other Hydrostatick experiments which he would often be so intent upon as to forget his dinner." If his mother gave him orders—to watch the sheep, "or upon any other rural employment"—as often as not Newton ignored her. Rather, "his chief delight was to sit under a tree with a book in his hands." Meanwhile, the flock wandered off or the pigs nosed into his neighbors' grain.

Hannah's attempt to break Newton to rural harness lasted nine months. He owed his escape to two men: his uncle, a clergyman and a graduate of Cambridge, and his former schoolmaster, William Stokes, who pleaded with Newton's mother to send her son to university. Hannah relented only when Stokes promised to pay the forty-shilling fee levied on boys born more than a mile from Cambridge.

Newton wasted no time getting out of town. Although the term would not begin until September, he set out from Woolsthorpe on June 2, 1661. He took almost nothing with him, and on arrival he equipped himself with a washstand, a chamber pot, a quart bottle, and "ink to fille it." Thus armed, Isaac Newton took up residence in Trinity, where he would remain for thirty-five years.

At Cambridge, it was Newton's ill luck to be poor—or rather, to be made so by Hannah, who again registered her disdain for book learning by limiting his allowance at university to ten pounds a year. That was not enough to cover food, lodging, and tutors' fees, so Newton entered Trinity as a subsizar—the name Cambridge gave to those students who paid their way by doing the tasks that the sons of richer men would not do for themselves. Having just left a prosperous farm with servants of his own, Newton was now expected to wait on fellow students at table, to eat their scraps, to haul wood for their fires, to empty pots filled with their piss.

Newton was not the most wretched among his fellow sizars. His ten-pound stipend counted for something, and he had a family connection to a senior member of the college. He could afford at least a few creature comforts. Cherries and marmalade show up in his expenses, as do such essentials as milk and cheese, butter and beer. But in his first years at the college, Newton lived at the very bottom of Trinity's hierarchy, standing while others sat, a man of no social consequence. He made almost no impression on the undergraduate life there. His entire correspondence contains just one letter to a college contemporary, written in 1669, five years after he completed his B.A. As Richard Westfall, Newton's leading biographer, has established, even after Newton became by far the most famous of his generation at Cambridge, not one of the students from his year admitted having met him.

There is no direct evidence to tell what Newton felt as he endured such solitude. But he did leave a powerful hint. In a notebook otherwise filled with expense records and geometry notes, he covered several pages in 1662 with what reads like a debtor's ledger of sins, entry after entry of transgressions large and small, a reckoning of the burden of debt owed to an unforgiving divine banker.

He admitted wrongs done to his fellow man: "Stealing cherry cobs from Eduard Storer / Denying that I did so"; "Robbing my mothers box of plums and sugar"; "Calling Derothy Rose a jade." He revealed an impressive urge to violence: "Punching my sister"; "Striking many"; "Wishing death and hoping it to some"; and in a brutal comment on his mother's remarriage, "Threatening my father and mother Smith to burne them and the house over them."

He admitted to gluttony, twice, and once, "Striving to cheat with a brass halfe crowne"—with hindsight, quite an admission for the man who would become the counterfeiters' scourge. He confessed to an escalating litany of crimes against God, petty misdemeanors like "Squirting water on Thy day" or "Making pies on Sunday night"; and then an agonized confession of mortal failure: "Not turning nearer to Thee according to my belief"; "Not Loving Thee for Thy self"; "Fearing man above Thee."

Worst of all, number twenty on his tally of fifty-eight failings convicted him of "Setting my heart on money learning pleasure more than Thee." Since the Temptation, money and the delights of the senses have been Satan's lures for the pious. But for Newton the true danger came from the snare that had captured Eve: an idolatrous love of knowledge. Trinity opened to Newton a world of ideas that had been closed to him in the countryside, and he entered it with ferocious concentration, so deep, it seems, that it drove God from his mind and heart.

Even at Cambridge, though, Newton had to find his own way. He recognized quickly that the traditional university curriculum, centered on Aristotle as the ultimate authority, was a waste of his time. His reading notes show that he never bothered to wade all the way through any of the assigned Aristotelian texts. Instead, Newton set himself to master the new knowledge that was trickling into Cambridge past the defenses of ancient authority. He did so mostly on his own—he had to, for his understanding soon surpassed that of all but one or two of the men on the faculty who could have instructed him.

He began with a glance at Euclid's geometry, but on first reading found its claims "so easy to understand that he wondered how any body would amuse themselves to write any demonstrations of them." More mathematics followed, and then he discovered mechanical philosophy—the notion that the entire material world could be understood as patterns of matter in motion. It was a controversial idea, mostly because it seemed, to some at least, to diminish the significance of God in daily life. But even so, Descartes and Galileo—and many others—had demonstrated the effectiveness of the new approach, to the point where the mechanical worldview reached all the way to the few receptive minds to be found in that backwater of European intellectual life, the University of Cambridge.

Newton's legendary capacity for study displayed itself here, in this first rush to master all that Europe knew of how the material world works. Sleep was optional. John Wickens, who arrived at Cambridge eighteen months after Newton, remembered that when Newton was immersed in his work, he simply did without. Food was fuel—and, as often as not, merely a distraction. He later told his niece that his cat grew fat on the meals he forgot to eat.

In 1664, after two hard years, Newton paused to sum up his learning in a document he modestly called *Quæstiones quædam Philosophicæ*—Certain Philosophical Questions. He started by asking what was the first or most basic form of matter, and in a detailed analysis argued that it had to be those simple, indivisible entities dubbed atoms. He posed questions on the true meaning of position—location in space—and of time, and of the behavior of celestial bodies. He probed his new and temporary master, Descartes, challenging his theory of light, his physics, his ideas about the tides. He sought to grasp how the senses worked. He had purchased a prism at the Sturbridge Fair in 1663, and now wrote up his first optical experiments, the starting point for his analysis of light and color. He wondered about motion and why a falling body falls, though he was confused about the property called gravity. He attempted to understand what it might mean to live in a truly mechanical universe, one in which all of nature except mind and spirit formed a grand and complicated machine—and then he trembled at the fate of God in such a cosmos. He wrote that "tis a contradiction to say ye first matter depends on some other subject." He added "except God"—and then crossed out those last two words.

He offered no definitive answers. This was the work of an apprentice mastering his tools. But it is all there in embryo, the program that would lead Newton toward his own discoveries and to the invention of the method that others could use to discover yet more. And while the Newtonian synthesis was decades away from completion, the *Quæstiones* captures the extraordinary ambition of an anonymous student working on the fringes of the learned world, who nonetheless proclaims his own authority, independent of Aristotle, of Descartes, of anyone.

Newton was fearless in the pursuit of anything he wanted to know. To find out whether the eye could be tricked into seeing what wasn't there, he stared directly at the sun through one eye for as long as he could bear the pain, then noted how long it took to free his sight from the "strong phantasie" of the image. A year or so later, when he wanted to understand the effect of the shape of an optical system on the perception of color, he inserted a bodkin—a blunt needle—"betwixt my eye and ye bone as near to ye backside of my eye as I could." Next, by "pressing my eye wth ye end of it (so as to make ye curvature . . . in my eye)" he saw several "white dark and coloured circles"—patterns that became clearer when he rubbed his eye with the point of his needle. To that description Newton helpfully added a drawing of the experiment, showing how the bodkin deformed his eye. It is impossible to look at the illustration without wincing, but Newton makes no mention of pain, nor any sense of danger. He had a question and the means to answer it. The next step was obvious.

He pressed on, pondering the nature of air, wondering whether fire could burn in a vacuum, taking notes on the motion of comets, considering the mystery of memory and the strange and paradoxical relationship of the soul to the brain. But, caught up as he was in the whirlwind of new thoughts, new ideas, he still had to deal with the ordinary obstacles of university life. In the spring of 1664, he sat for the one examination required of undergraduates at Cambridge, a test that would determine whether he would become one of Trinity's scholars. Pass, and he would cease to be a sizar; the college would pay his board and give him a small stipend for the four years it would take to become master of arts. Fail, and it was back to the farm.

He survived the ordeal, receiving his scholarship on April 28, 1664. But his renewed studies at the college were interrupted within months. Early in 1665, rats turned up on the docks along the Thames which had almost certainly come by way of Holland, perhaps in ships carrying prisoners from the Dutch wars or smuggled bales of cotton from the Continent. The rats carried their own cargo of fleas across the North Sea, and the fleas in turn ferried into England the bacterium *Yersinia pestis*. The fleas leapt from the rats; they bit; the bacteria slid into human veins, and dark buboes began to sprout. The bubonic plague had returned to England.

At first the disease proceeded slowly, a troubling backdrop to the daily routine. The first named victim died on April 12 and was buried in haste that same day in Covent Garden. Samuel Pepys noted "Great fears of the Sicknesse" in his diary entry for April 30. But the great naval victory over the Dutch at Lowestoft distracted him and many others. Then, in early June, Pepys found himself, "much against my Will," walking in Drury Lane, where he saw "two or three houses marked with a red cross upon the doors and 'Lord have mercy upon us' writ there."

That day, Pepys bought a roll of tobacco to chew, "which took away the apprehension." But the epidemic had taken hold, and no amount of nicotine could hold back panic. A thousand a week died in London, then two, until by September the death toll reached one thousand each *day*.

The very concept of a funeral collapsed under the weight of corpses. The best that could be done was disposal, landfill. As Daniel Defoe described it: A death cart enters a cemetery, halting at a broad pit. A man follows, walking behind the remains of his family. And then, "no sooner was the cart turned round and the bodies shot into the pit promiscuously, which was a surprise to him," Defoe wrote, "for he at least expected they would have been decently laid in." Instead, "Sixteen or seventeen bodies; some were wrapt up in linen sheets, some in rags, some little other than naked, or so loose that what covering they had fell from them in the shooting out of the cart, and they fell quite naked among the rest; but the matter was not much to them, or the indecency much to any one else, seeing they were all dead, and were to be huddled together into the common grave of mankind." This was democracy at last, "for here was no difference made, but poor and rich went together; there was no other way of burials, neither was it possible there should, for coffins were not to be had for the prodigious numbers that fell in such a calamity as this."

Those who could fled as fast as possible, but the disease ran with the refugees, and the dread of the plague reached farther and farther into the countryside. Cambridge emptied early, becoming a ghost town by midsummer 1665. The great fair at Sturbridge—England's largest—was canceled. The university ceased to offer sermons in Great St. Mary's Church, and on August 7, Trinity College acknowledged the obvious by authorizing the payment of stipends to "all Fellows & Scholars which now go into the Country on occasion of the Pestilence."

Newton was already long gone, escaping before the August stipend came due. He retreated to Woolsthorpe, its isolation a sanctuary from any chance encounter with a plague rat or a diseased person. He seems not to have noticed the change of scene. No one now dared set the prodigal to the plow. In the last months before Newton abandoned Cambridge, his mind had turned almost exclusively to mathematics. In the quiet of his home, he continued, building the structure that would ultimately revolutionize the mathematical understanding of change over time. Later in the plague season, he would take the first steps toward his theory of gravity, and thereby toward his understanding of what governs motion throughout the cosmos.

The disease cut through England all that summer and fall, murdering its tens of thousands. Isaac Newton paid it little mind. He was busy.

2

"The Prime of My Age"

THE PLAGUE OF 1665 raged on through the fall. In December, a bitter cold settled across the south of England. Samuel Pepys wrote that the hard frost "gives us hope for a perfect cure of the plague." But the disease persisted—up to thirteen hundred Londoners a week were still dying—and prudent folk shunned crowds if they could.

Isaac Newton was cautious to a fault. He celebrated his twenty-third birthday that Christmas Day at home, safely distant from the infectious towns. He stayed there into the new year, working, he said, with an intensity he never again equaled: "In those days," he remembered fifty years on, "I was in the prime of my age for invention & minded Mathematicks & Philosophy more then at any time since."

Mathematics first, continuing what he had started before his enforced retreat from Cambridge. The critical ideas emerged from the strange concept of the infinite, in both its infinitely large and infinitesimally small forms. Newton would later name the central discovery of that first plague year the "method of fluxions." In its developed form, we now call it the calculus, and it remains the essential tool used to analyze change over time.

He did not complete this work in total isolation. In the midst of his thinking about infinitesimals, the epidemic seemed to ease in the east of England. By March, Cambridge town had been free of plague deaths for six consecutive weeks. The university reopened, and Newton returned to Trinity College. In June, though, the disease reappeared, and on the news of more deaths,

Newton again fled home to Woolsthorpe. Back on the farm, his attention shifted from mathematics to the question of gravity.

The word already had multiple meanings. It could imply a ponderousness of spirit or matter—the affairs of nations had gravity, and to be said to possess gravitas was a badge of honor for the leaders of nations. It had a physical meaning too, but what it was—whether a property of heavy objects or some disembodied agent that could act on objects—no one knew. In the *Quæstiones,* Newton had titled one essay "Of Gravity and Levity," and he wrestled there with concepts that he found to be vague and indistinct. He wrote of "the matter causing gravity" and suggested that it must pass both into and out of "the bowels of the earth." He considered the question of a falling body and wrote of "the force which it receives every moment from its gravity"—that is, force somehow inherent in the object plummeting toward the ground. He wondered whether "the rays of gravity may be stopped by reflecting or refracting them." For the time being, all that Newton knew about the connection between matter and motion was that one existed.

Now, in his enforced seclusion, Newton tried again. According to legend, the key idea came to him in one blinding flash of insight. Sometime during the summer of 1666, he found himself in the garden at Woolsthorpe, sitting "in a contemplative mood," as he remembered—or perhaps invented, recalling the moment decades later, in the grip of nostalgia and old age. In his mind's eye the apple tree of his childhood was heavy with fruit. An apple fell. It caught his attention. Why should that apple always descend perpendicularly to the ground, he asked himself. Why should it go not sideways or upward but constantly to the earth's center?

Why not indeed. The myth that has endured from that time to this declared that that was all it took: on the spot, Newton made the leap of reason that would lead to the ultimate prize, his theory of gravity. Matter attracts matter, in proportion to the mass contained in each body; the attraction is to the center of a given mass; and the power "like that we here call gravity ... extends its self thro' the universe."

Thus the story of what one author has called the most significant apple since Eve's. It has the virtue of possessing some residue of fact. The tree itself existed. After his death, the original at Woolsthorpe was still known in the neighborhood as Sir Isaac's tree, and every effort was made to preserve it, propping up its sagging limbs until it finally collapsed in a windstorm in 1819. A sliver of the tree ended up at the Royal Astronomical Society, and branches had already been grafted onto younger hosts, which in time bore fruit of their own. In 1943, at a dinner party at the Royal Society Club, a member pulled from his pocket two large apples of a variety called Flower of Kent, a cooking apple popular in the 1600s. These were, the owner explained, the fruit of one of the grafts of the original at Woolsthorpe. Newton's apple itself is no fairy tale; it budded, it ripened; almost three centuries later it could still be tasted in all the knowledge that flowed from its rumored fall.

But whatever epiphany Newton may have had in that plague summer, it did not include a finished theory of gravity. At most, the descent of that apple stimulated the first step in a much longer, more difficult, and ultimately much more impressive odyssey of mental struggle, one that took Newton from concepts not yet formed all the way to a finished, dynamic cosmology, a theory that reaches across the entire universe.

That first step, of necessity, turned on the existing state of knowledge, both Newton's and that of European natural philosophers. Earlier in the plague season, Newton had studied how an object moving in a circular path pushes outward, trying to recede from the center of that circle—a phenomenon familiar to any child twirling a stone in a sling. After a false start, he worked out the formula that measures that centrifugal force, as Newton's older contemporary, Christiaan Huygens, would name it. This was a case of independent invention. Huygens anticipated Newton but did not publish his result until 1673. That is: Newton, just twenty-two, was working on the bleeding edge of contemporary knowledge. Now to push further.

He did so by testing his new mathematical treatment of circular motion on the revolutionary claim that the earth did not stand still at the center of a revolving cosmos. One of the most potent objections to Copernicus's sun-centered system argued that if the earth really moved around the sun, turning on its axis every day as it went, that rotation would generate so much centrifugal force that humankind and everything else on the surface of such an absurdly spinning planet would fly off into the void. With his new insight, however, Newton realized that his formula allowed him to determine just how strong this force would be at the surface of the turning earth.

To begin, he used a rough estimate for the earth's size—a number refined over the previous two centuries of European exploration by sea. With that, he could figure the outward acceleration experienced at the surface of a revolving earth. Next, he set out to calculate the downward pull at the earth's surface of what he called gravity, in something like the modern sense of the term. Galileo had already observed the acceleration of falling bodies, but Newton trusted no measurement so well as one he made himself, so he performed his own investigation of falling objects by studying the motion of a pendulum. With these two essential numbers, he found that the effect of gravity holding each of us down is approximately three hundred times stronger than the centrifugal push urging us to take flight.

It was a bravura demonstration, an analysis that would have placed Newton in the vanguard of European natural philosophy, had he told anyone about it. Even better, he found he could apply this reasoning to a larger problem, the behavior of the solar system itself. What was required, for example, to keep the moon securely on its regular path around the earth? Newton knew one fact: any such force would strain against the moon's centrifugal tendency to recede, to fly off, abandoning its terrestrial master. At the appropriate distance, he realized, those impulses must balance, leaving the moon to fall forever as it followed its (nearly) circular path around the center of the earth, the source of that still mysterious impulse that would come to be called gravity.

Mysterious, but calculable. To do so, he needed to take one last, great step and create a mathematical expression to describe the intensity of whatever it was connecting the earth and the moon with the distance between the two bodies. He found inspiration in Kepler's third law of planetary motion, which relates the time it takes for a planet to complete its orbit with its distance from the sun. By analyzing that law, Newton concluded (as he later put it) that "the forces which keep the planets in their orbs must [be] reciprocally as the squares of their distances from the centers around which they revolve." That is, the force of gravity falls off in proportion to the square of the distance between any two objects.

With that, it was just a matter of plugging in the numbers to calculate the moon's orbit. Here he ran into trouble. From his pendulum experiments, he had a fairly precise measurement of one crucial term, the strength of gravity at the earth's surface. But he still needed to know the distance between the moon and the earth, a calculation that turned on knowledge of the earth's size. This was a number Newton could not determine for himself, so he used the common mariner's guess that one degree of the earth's circumference was equal to "sixty measured Miles only." That was wrong, well off the accurate figure of slightly more than sixty-nine miles. The error propagated throughout his calculation, and nothing Newton could do would make the moon's path work out. He had some guesses as to what might be happening, but these were loose thoughts, and as yet he knew no way to reduce them to the discipline of mathematics.

The setback was enough to provoke Newton to move on. New ideas were crowding in. Optics came next, a series of inquiries into the nature of light that would bring him a first, ambivalent brush with fame in the early 1670s. Thus engaged, Newton let the matter of the moon rest.

But if his miracle years, as they have come to be known, did not produce the finished Newtonian system, still by the end of his enforced seclusion Newton understood that any new physical system could succeed only by "subjecting motion to number." His attempt to analyze the gravitational interaction of the earth and the moon provided the model: any claim of a relationship, any proposed connection between phenomena, had to be tested against the rigor of a mathematical description.

Many of the central ideas that would form the essential content of his physics were there too, though an enormous amount of labor remained to get from those first drafts to the finished construction of the system. Newton would have to redefine what he and his contemporaries thought they knew about the most basic concepts of matter and motion just to arrive at a set of definitions that he could turn to account. For example, he was still groping for a way to express the crucial conception of force that would allow him to bring the full force of mathematics to bear. By 1666, he had got this far: "Tis known by ye light of nature ... y^t equall forces shall effect an equall change in equall bodys ... for in loosing or ... getting ye same quantity of motion a body suffers the ye same quantity of mutation in its state."

The core of the idea is there: that a change in the motion of a body is proportional to the amount of force impressed on it. But to turn that conception into the detailed, rich form it would take as Newton's second law of motion would require long, long hours of deep thought. The same would prove to be true for all his efforts over the next twenty years as they evolved into the finished edifice of his great work, *Philosophiæ naturalis principia mathematica*—The Mathematical Principles of Natural Philosophy—better known as the *Principia*. For all his raw intelligence, Newton's ultimate achievement turned on his genius for perseverance. His one close college friend, John Wickens, marveled at his ability to forget all else in the rapt observation of the comet of 1664. Two decades later, Humphrey Newton, Isaac Newton's assistant and copyist (and no relation), saw the same. "When he has sometimes taken a turn or two [outdoors] has made a sudden stand, turn'd himself about and run up ye stairs, like another Archimedes, with an Eureka, fall to write on his Desk standing, without giving himself the Leasure to draw a Chair to sit down in." If something mattered to him, the man pursued it relentlessly.

Equally crucial to his ultimate success, Newton was never a purely abstract thinker. He gained his central insight into the concept of force from evidence "known by ye light of nature." He tested his ideas about gravity and the motion of the moon with data drawn from his own painstaking experiments and the imperfect observations of others. When it came time to analyze the physics of the tides, the landlocked Newton sought out data from travelers the world over; barely straying from his desk in the room next to Trinity College's Great Gate, he gathered evidence from Plymouth and Chepstow, from the Strait of Magellan, from the South China Sea. He stabbed his own eye, built his own furnaces, constructed his own optical instruments (most famously the first reflecting telescope); he weighed, measured, tested, smelled, worked—hard—with his own hands, to discover the answer to whatever had sparked his curiosity.

Newton labored through the summer. That September, the Great Fire of London came. It lasted five days, finally exhausting itself on September 7. Almost all of the city within the walls was destroyed, and some beyond, 436 acres in all. More than thirteen thousand houses burned, eighty-seven churches, and old St. Paul's Cathedral. The sixty tons of lead in the cathedral roof melted; a river of molten metal flowed into the Thames. Just six people are known to have died, though it seems almost certain that the true number was much greater.

But once the fire destroyed the dense and deadly slums that cosseted infection, the plague finally burned itself out. That winter, reports of cases dropped, then vanished, until by spring it became clear that the epidemic was truly done.

In April 1667, Newton returned to his rooms at Trinity College. He had left two years earlier with the ink barely dry on his bachelor of arts degree. In the interval, he had become the greatest mathematician in the world, and the equal of any natural philosopher then living. No one knew. He had published nothing, communicated his results to no one. So the situation would remain, in essence, for two decades.

3

"I Have Calculated It"

ISAAC NEWTON CLAMBERED up the academic pyramid as rapidly as his abilities warranted. In 1669, when Newton was twenty-six, his former teacher Isaac Barrow resigned the Lucasian Professorship of Mathematics in his favor, and from that point on he was set. The chair was his for as long as he chose to keep it. It provided him with room, board, and about one hundred pounds a year—plenty for an unmarried man with virtually no living expenses. In return, all he had to do was deliver one course of lectures every three terms. Even that duty did not impinge much on his time. Humphrey Newton reported that the professor would speak for as much as half an hour if anyone actually showed up, but that "oftimes he did in a manner, for want of Hearers, read to ye Walls."

Aside from such minimal nods toward the instruction of the young, Newton did as he pleased. He loathed distractions, had little gift for casual talk, and entertained few visitors. He gave virtually all his waking hours to his research. Humphrey Newton again: "I never knew him [to] take any Recreation or Pastime, either in riding out to take air, Walking, bowling, or any other Exercise whatever, Thinking all Hours lost, that was not spent in his Studyes." He seemed offended by the demands of his body. Humphrey reported that Newton "grudg'd that short Time he spent in eating & sleeping"; that his housekeeper would find "both Dinner & Supper scarcely tasted of"; that "He very seldom sat by the fire in his Chamber, excepting that long frosty winter, which made him creep to it against his will." His one diversion was his garden, a small plot on Trinity's grounds, "which was never out of Order, in which he would, at some seldom Times, take a short Walk or two, not enduring to see a weed in it." That was it—a life wholly committed to his studies, except for a very occasional conversation with a handful of acquaintances and a few stolen minutes pulling weeds.

But work to what end? Year after year, he published next to nothing, and he had almost no discernible impact on his contemporaries. As Richard Westfall put it: "Had Newton died in 1684 and his papers survived, we would know from them that a genius had lived. Instead of hailing him as a figure who had shaped the modern intellect, however, we would at most ... [lament] his failure to reach fulfillment."

And then, one August day in 1684, Edmond Halley stopped by. Halley was one of that handful of acquaintances who could always gain admittance to Newton's rooms in Trinity. The pair had met two years earlier, just after Halley's return from France, where he'd meticulously observed the comet that would later be named for him. Newton had made his own sketches of the comet, and he welcomed a fellow enthusiast into the circle of those whose letters he would answer, whose conversation he welcomed.

Today Halley brought no pressing scientific news. He had come down from London to the countryside near Cambridge on family business, and his visit to Newton was merely social. But in the course of their conversation, Halley recalled a technical point he had been meaning to take up with his friend.

Halley's request had seemed trivial enough. Would Isaac Newton please settle a bet? The previous January, Halley, Robert Hooke, and the architect Sir Christopher Wren had talked on after a meeting of the Royal Society. Wren wondered if it was true that the motion of the planets obeyed an inverse square law of gravity—the same inverse square relationship that Newton had investigated during the plague years. Halley readily confessed that he could not solve the problem, but Hooke had boasted that he had already proved that the inverse square law held true, and "that upon that principle all the Laws of the celestiall motions were to be demonstrated."

When pressed, though, Hooke refused to reveal his results, and Wren openly doubted his claim. Wren knew how tricky the question was. Seven years before, Isaac Newton had visited him in his London home, where the two men discussed the complexity of the problem of discovering "heavenly motions upon philosophical principles." Accordingly, Wren would not take a claim of a solution on faith. Instead, he offered a prize, a book worth forty shillings, to the man who could solve the problem within two months. Hooke puffed up, declaring that he would hold his work back so that "others triing and failing, might know how to value it." But two months passed, and then several weeks more, and Hooke revealed nothing. Halley, diplomatically, did not write that Hooke had failed, but that "I do not yet find that in that particular he has been as good as his word."

There the matter rested, until Halley put Wren's question to Newton: "what he thought the Curve would be that would be described by the Planets supposing the force of attraction towards the Sun to be reciprocal to the square of their distance from it." Newton immediately replied that it would be an ellipse. Halley, "struck with joy & amazement," asked how he could be so sure, and Newton replied, "Why . . . I have calculated it."

Halley asked at once to see the calculation, but, according to the story he later told, Newton could not find it when he rummaged through his papers. Giving up for the moment, he promised Halley that he would "renew it & send it to him."

While Halley waited in London, Newton tried to re-create his old work—and failed. He had made an error in one of his diagrams in the prior attempt, and his elegant geometric argument collapsed with the mistake. He labored on, however, and by November he had worked it out.

In his new calculation, Newton analyzed the motions of the planets using a branch of geometry concerned with conic sections. Conic sections are the curves made when a plane slices through a cone. Depending on the angle and location of the cut, you get a circle (if the plane intersects either cone at a ninety-degree angle), an ellipse (if the plane bisects one cone at an angle other than ninety degrees), a parabola (if the curve cuts through the side of the cones but does not slice all the way through its circumference), or the symmetrical double curve called a hyperbola (produced only if there are two identical cones laid tip to tip).

As he calculated, Newton was able to show that for an object in a system of two bodies bound by an inverse square attraction, the only closed path available is an ellipse, with the more massive body at one focus. Depending on the distance, the speed, and the ratio of masses of the two bodies, such ellipses can be very nearly circular—as is the case for the earth, whose orbit deviates by less than two percent from a perfect circle. As the force acting on two bodies weakens with distance, more elongated ellipses and open-ended trajectories (parabolas or hyperbolas) become valid solutions for the equations of motion that describe the path of a body moving under the influence of an inverse square force. To the practical matter at hand, Newton had proved that in the case of two bodies, one orbiting the other, an inverse square relationship for the attraction of gravity produces an orbit that traces a conic section, which becomes the closed path of an ellipse in the case of our sun's planets.

QED.

Newton wrote up the work in a nine-page manuscript titled *De motu corporum in gyrum*—"On the Motion of Bodies in Orbit." He let Halley know the work was done, and then presumably settled back into his usual routine.

That peace could not last, not if Halley had anything to do with it. He grasped the significance of *De Motu* immediately. This was no mere set-piece response to an after-dinner challenge. Rather, it was the foundation of a revolution of the entire science of motion. He raced back to Cambridge in November, copied Newton's paper in his own hand, and in December was able to tell the Royal Society that he had permission to publish the work in the register of the Royal Society as soon as Newton revised it.

And then . . . nothing.

Halley had not expected anything more than a quick revision of the brief paper he had already seen. The final, corrected version of *De Motu* was supposed to follow soon after his second meeting with Newton. When it failed to arrive on schedule, Halley took the precaution of registering his preliminary copy with the Royal Society, establishing its priority. Then he resumed his vigil, waiting for more to come from Cambridge. Still nothing, not in what remained of 1684, and not through the first part of 1685.

Newton, for all of his periodic public silences, wrote constantly. He committed millions of words to paper over his long life, often recopying three or more near-identical drafts of the same document. He was a conscientious letter writer too. His correspondence fills seven folio volumes. While that is not an extraordinary total for a time when the learned of Europe (and America) communicated with each other by letter, it represents a formidable stream of prose. But between December 1684 and the summer of 1686, when he delivered to Halley the final versions of the first two parts of his promised, and now greatly expanded, treatise, he is known to have written just seven letters.

Two of them are mere notes. The remaining five were all to John Flamsteed, the Astronomer Royal, asking him for his observations of the planets, of Jupiter's moons, and of comets, all to help him in a series of calculations whose true nature he did not choose to share.

Much later, Newton admitted what had happened. "After I began to work on the inequalities of the motions of the moon, and then also began to explore other aspects of the laws and measures of gravity and of other forces," he wrote, "I thought that publication should be put off to another time, so that I might investigate these other things and publish all my results together." He was trying to create a new science, one he called "rational mechanics." This new discipline would be comprehensive, able to gather in the whole of nature. It would be, he wrote, "the science, expressed in exact propositions and demonstrations, of the motions that result from any forces whatever and of the forces that are required for any motions whatever."

Newton writes here of a science advanced by a method that would be exact in its laws and analyses. Fully developed, it would yield an absolute, precise account of cause and effect, true for all encounters between matter and force, whatever they may be. This was his aim in writing what was about to become the *Principia,* at once the blueprint and the manifesto for such a science. He began with three simple statements that could cut through the confusion and muddled thought that had tangled all previous attempts to account for motion in nature. First came his ultimate understanding of what he dubbed inertia: "*Every body perseveres in its state of being at rest or of moving uniformly straight forward except insofar as it is compelled to change its state by forces impressed.*"

His second axiom stated the precise relationship between force and motion: "*A change in motion is proportional to the motive force impressed and takes place along the straight line in which that force is impressed.*" Last he addressed the question of what happens when forces and objects interact: "*To any action there is always an opposite and equal reaction; in other words, the actions of two bodies upon each other are always equal and always opposite in direction*" (italics in the original).

Thus, the famous three laws of motion, stated not as propositions to be demonstrated but as pillars of reality. This was, Newton recognized, an extraordinary moment, and he composed his text accordingly, in an echo of the literature he knew best. He began with a revelation, a bald statement of fundamental truths, then followed with five hundred pages of exegesis that showed what could be done from this seemingly simple point of origin.

Books One and Two—both titled "The Motion of Bodies"—demonstrated how much his three laws could explain. After some preliminaries, Newton reworked the material he had shown Halley to derive the properties of the different orbits produced by an inverse square law of gravity. He analyzed mathematically how objects governed by the three laws collide and rebound. He calculated what happens when objects travel through different media—water instead of air, for example. He pondered the issues of density and compression, and created the mathematical tools to describe what happens to fluids under pressure. He analyzed the motion of a pendulum. He inserted some older mathematical work on conic sections, apparently simply because he had it lying around. He attempted an analysis of wave dynamics and the propagation of sound. On and on, through every phenomenon that could be conceived as matter in motion.

He wrote on through the fall and winter of 1685, stating propositions and theorems, presenting proofs, extracting corollaries from concepts already established, page after page, proof after proof, until the sheer mass overwhelmed all challenges. Throughout that time, Newton's always impressive appetite for work became total. "He very rarely went to Bed till 2 or 3 of the clock, sometimes not till 5 or 6, lying about 4 or 5 hours," observed Humphrey Newton. On rising, "his earnest & indefatigable Studyes retain'd Him, so that He scarcely knewe the Hour of Prayer."

It took Newton almost two years to finish Book Two. Its last theorem completes his demolition of Descartes' vortices—those whirlpools in some strange medium that were supposed to drive the motion of the planets and stars. Newton showed no pity, concluding dismissively that his predecessor's work served "less to clarify the celestial motions than to obscure them."

With that bit of old business settled, Newton turned to his ultimate aim. In the preface to the *Principia*, Newton wrote, "The whole difficulty of philosophy seems to be to discover the forces of nature from the phenomena of motions and then to demonstrate the other phenomena from these forces." Books One and Two had covered only the first half of that territory, presenting "the laws and conditions of motions"; but as Newton wrote, those laws were "not, however, philosophical but strictly mathematical." Now, he declared, it was time to put such abstraction to the test of experience. "It still remains for us," he wrote, "to exhibit the system of the world from these same principles."

At first reading, Book Three, which he in fact titled "The System of the World," falls short. No mere forty-two propositions could possibly comprehend all of experience. But, as usual, Newton said what he meant. In a mere hundred pages or so of mathematical reasoning, he did not promise to capture all that moved in the observable universe. Rather, he offered a system with which to do so—the method that, as it has turned out, his successors have employed to explore all of material reality through the enterprise we call science.

As Book Three opens, gravity at last takes over the entire narrative. Once again, Newton begins with the foundational claims of his investigation. Most important, he states what can be seen as the fundamental axiom of science: that the properties of objects that can be observed on earth must be assumed to be properties of bodies anywhere in the cosmos. Here he demonstrates that gravity behaves the same way whether it pulls a cannonball back to the ground or tugs on the most distant object in the heavens. He shows that the satellites of Jupiter obey his inverse square law of gravitation, then runs through the same reasoning for the major planets and for the moon.

Next he proves that the center of the planetary system must be the sun, and explores how the mutual gravitational attraction between Saturn and Jupiter pulls both planets' orbits away from the perfect ellipse of a geometer's dream. Mathematics, Newton here affirms, is essential for the analysis of the physical world, but nature itself is more complex than any purely mathematical idealization of it.

Newton races on—so many phenomena, only so much time and energy with which to explore them. Closer to home, he analyzes the track of the moon and the implications of the observed fact that the earth is not a perfect sphere. (He proved that the gravitational pull of a spheroid would not be the same everywhere, and hence one's weight would vary slightly depending on where one stands on the earth's surface.) And, seemingly at the end of a journey from the outermost known planets to the surface of the earth, he examines the influence of moon and sun on the earth's tides. Twenty years after he looked at gravity as a purely local phenomenon, Newton here presents gravity as the engine of the system of all creation—one that binds the rise and fall of the Thames or the Gulf of Tonkin to all the observed motions of the solar system.

But Newton does not choose to end Book Three here, and his decision reveals how much the work as a whole acts to persuade and not merely to demonstrate. To be sure, no one thinks of Newton as a novelist, or of the *Principia* as a galloping read. But Book Three—and the volume in its entirety—can be experienced as a kind of epic of gravity, and to bring that tale to its heroic close, Newton spins his account outward once again, into the realm of the comets.

The passage begins slowly, with a detailed, tedious series of observations of the path of the Great Comet of 1680, the product of Newton's relentless attempts to distinguish good data from bad. From that base of unassailable evidence, Newton plots an orbit. Then he derives the same path by calculation, extracting the comet's course from just three observed positions. The two tracks—the one observed and the one predicted—match almost exactly, tracing the curve called a parabola. It does not take a huge change in trajectory to place a comet on a parabolic path instead of an elliptical one, but the distinction is crucial. Comets in elliptical orbit, like that of 1682, which we now call Halley's, return again and again. A comet on a parabolic journey passes near the earth just once. It swings by the sun and then keeps going, traveling on a path that can, in principle, carry it to the farthest extremes of the heavens.

With this, the *Principia* reaches its true climax. Nothing in Newton's science depends on the shape of this narrative. In any order, his proofs would be just as valid. But to take the reader on an odyssey that begins with the orbits of the planets and extends to bring the entire cosmos into view allows the larger implications of the Newtonian idea to emerge. At the end of the discussion of the comet of 1680, he writes, "The theory that corresponds exactly to so nonuniform a motion through the greatest part of the heavens, and that observes the same laws as the theory of the planets and that agrees exactly with exact astronomical observations *cannot fail to be true*" (italics added).

Truth, omnipresent and omnipotent: the *Principia* reveals laws of motion and gravity that do not merely describe how cannonballs fly or apples fall; they do not simply hold the earth in its orbit around the sun or regulate the dance of Saturn's moons around the ringed planet. Instead, as promised, Newton offered his world an idea that encompasses all matter, all motion, to the deepest reaches of the imaginable universe, a cosmos mapped by the paths of comets tracing out their elegant curves in journeys that end at infinity.

And then Newton rested. Edmond Halley received Book Three of the *Principia* on April 4, 1687. He spent the next three months in publisher's hell. He split the printing job between two shops, whose work had to be coordinated and supervised. Between the mathematical formulas and the woodcut illustrations, some of the sheets were so complicated that Halley found himself consumed by the demands of the book. He confided to a friend that "Mr. Newton's book . . . has made me forget my duty in regard of the Societies correspondents," and that "the correction of the press costs me a great deal of time and pains." He never complained openly to Newton himself, however, writing instead of "your divine treatise" and "your excellent work."

Halley ordered a run of between two hundred fifty and four hundred copies from the printers. The finished books arrived on July 5, 1687. Halley sent twenty copies to Newton. Most of the rest went on sale. At seven shillings apiece unbound, two shillings more for a leather binding, the edition sold out almost immediately. Newton's life was about to be transformed.

4

"The Incomparable Mr. Newton"

FOR JOHN LOCKE, 1691 had been a busy year. He had left
London for an open-ended stay at a friend's country house in
Essex, and he had completed another book, one of his first since
A Letter Concerning Toleration, his famous argument for freedom of conscience and belief. The new work took on a completely different though equally contentious topic: what to do
about England's growing financial crisis, brought on by the
plague of bad coins. After sending friends copies of the new
manuscript in early December, he found himself free of immediate duties. So, at leisure at last, he resumed one of the hobbies
of his youth.

Just before nine o'clock on the morning of Sunday, December
13, he left his rooms upstairs, overlooking the garden, and hurried outside to record his daily observations of the weather. His
thermometer was a good one, produced by the celebrated London watchmaker Thomas Tompion. Locke recorded the tem-
perature: 3.4 on the particular scale used on his instrument—notably colder than the "temperate" reading of 4, but not quite as
cold as the day before, when Locke noted frost. This day, he
found that the barometric pressure had dropped overnight and a
light breeze had set in from the east. Last, he recorded the condition of the sky: thick, uniform clouds. In other words, a typical
December day in the east of England: chilly, damp, and dull.

That same day, about thirty miles to the north, Isaac Newton, in a state of annoyance, began a letter. He drew out a sheet
of paper, loaded his quill with ink, and began to write. He filled

a page, read it, and paused. Newton was swift to take offense, and as Robert Hooke had already learned to his sorrow, Newton's enemies had to expect overwhelming retaliation for any slight, real or imagined. But today's missive was directed against that amateur meteorologist John Locke, a man whom Newton admired and by whom he was admired in turn. Newton found it difficult to strike the right note of reproach.

The crime in question? Locke had offered to help his friend Newton gain the post of Master of Charterhouse, a boys' school in London. Newton recoiled at the thought. "You seem still to think on Charterhouse," he wrote, but "I believe your notions & mine are very different about the matter." What was wrong with the proposal? Everything. "The competition is hazzardous," he complained, "and I am loathe to sing a new song" in hopes of persuading the mighty to throw him a sop. Still more galling, the pay was meager, beneath him. "Its but 200 pounds per an besides a Coach (wch I reccon not) & lodging"—not enough to live in the style to which Newton aspired nor fit reward for a man of his reputation.

And, of course, there was the problem of London.

Newton had lived in Cambridge for thirty years. All the decades of thought and labor that had transformed an awkward country boy into the dominant mind in Europe had taken place in and around the rooms overlooking the Great Court and chapel of Trinity College, from which he now wrote angrily to his friend. And yet Locke dared to suggest that he should abandon Cambridge for London, with all its filth and pretense. How could Newton express the manifold unsuitability of the suggestion? Try this: "The confinement to ye London air & a formal way of life is what I am not fond of."

Line after line expressed his sense of insult—and then he stopped. His rage cooled. He did not sign the letter.

The truth was that Newton desperately hoped to escape his intellectual cloister, and just as desperately desired the exceptionally well-connected Locke's help to do so. What had happened?

The *Principia* had, and with it Newton's sudden emergence into the circles of the great.

From the moment of its publication—and before, in fact—Edmond Halley had done his best to make sure that the *Principia* received its proper reception. He launched his campaign on the first pages of the work itself, adding to Newton's text a dedicatory ode: "Error and doubt no longer encumber us with mist; / . . . We are now admitted to the banquets of the Gods; / We may deal with laws of heaven above; and we now have / The secret keys to unlock the obscure earth." And, lest anyone mistake the value of the man who had found the keys to the kingdom, Halley concluded: "Join me in singing the praises of NEWTON, who reveals all this / . . . No closer to the gods can any mortal rise." More soberly, in his formal review Halley argued for Newton's unique significance. "This incomparable author having at length been prevailed upon to appear in publick, has in this treatise given a most notable extent of the powers of the Mind." This Newton was the new Moses, a prophet revealing the law to the people: he had "at once shewn what are the principles of Natural Philosophy and so far derived from their consequences that he seems to have . . . left little to be done by those that shall succeed him."

Newton could, of course, count on Halley's praise. The reaction that truly mattered would come from the rest of learned Europe. Over the summer and into the autumn of 1687 those responses came in. *Acta Eruditorum,* Europe's leading scientific journal, called the book "an investigation worthy of so great a mathematician." In Paris, the devout Cartesian who reviewed the *Principia* for *Le Journal des sçavans* wanted an account of gravity that would reveal the mechanism by which one object attracted another, the kind of direct connection required by orthodox mechanical philosophers. The *Principia*'s purely mathematical description of gravity emphatically did not supply that kind of explanation, relying instead on the seemingly occult notion of forces acting across space—but the French reviewer still conceded that "it is not possible to make demonstrations more precise than those which [Newton] gives." The then-anonymous Scottish mathematician David Gregory wrote to Newton, offering "my most hearty thanks for having been at the pains to teach the world that which I never expected any man should have knowne." And though "your book is of so transcendent fineness and use that few will understand it," he stressed his awe on behalf of "those few who cannot but be infinitely thankful to you." Gottfried Leibniz was one of that little band who could indeed comprehend the work. His praise came in the most revealing form: in the winter of 1688–89 he rushed into print three articles that suggested he had either earlier arrived at or refuted some of Newton's conclusions. Such attempted theft acknowledged the obvious: the *Principia* had become the measure of all scientific excellence from the moment it appeared in print.

From there, it did not take long for Newton's fame to reach the next level. After discussing parts of the *Principia*, the French philosopher Marquis de l'Hospital burst out, "Good god what fund of knowledge there is in that book!" And then he pressed his companion, an acquaintance of Newton's, for "every particular of Sr I. even to the color of his hair [and] . . . does he eat and drink & sleep?" Then the Marquis asked the iconic question, the one that has chased Newton ever since: "Is he like other men?"

Newton had entered a realm of fame that catapulted him out of the narrow company of natural philosophers and into the wide world. One of the most worldly to fall into his orbit was an expatriate English man of letters living in the Netherlands—that genteel revolutionary John Locke. Late in 1687 Locke heard of a new book that was causing a sensation. He borrowed a copy from his friend Christiaan Huygens. But when Locke tried to read it, he found himself adrift in Newton's calculations. So he asked Huygens—after Newton the most important scientific thinker of the day—whether he could accept the *Principia*'s technical arguments on faith, simply assuming their validity. Huygens confirmed that Newton had proved what he had claimed, and so Locke read on, taking each mathematical conclusion for granted.

He was enthralled. He wrote one of the early, influential reviews of the book in 1688, in the *Bibliothèque universelle,* and he made sure his English readers took the point, writing in the preface to his *Essay on Human Understanding* in 1689 that "the commonwealth of learning is not at this time without masterbuilders, whose mighty designs, in advancing the sciences, will leave lasting monuments." Chief among them "the incomparable Mr. Newton." The critical Newtonian advance, Locke wrote, was that "we might in time hope to be furnished with more true and certain Knowledge in several parts of this stupendous Machine [Nature] than hitherto we could have expected."

Locke was eager to meet any man who had devised the path to such certain truth. There was just one problem: in 1687 he was a political exile, a wanted enemy of the English state. Four years before, Locke, thanks to his long association with King Charles II's Whig enemies, had been under routine surveillance by agents of the Crown when the Rye House Plot broke. The Rye House conspirators had planned to assassinate the King and his brother James, and the collapse of the scheme led to a wider roundup of the usual suspects. Several prominent Whigs were brought to trial and sent to the scaffold, and Locke himself faced arrest and possible execution for his guilt by association with one of the leading conspirators. Sensibly, he began to move around England and then fled the country altogether, reaching the Netherlands in September 1683. As long as the Stuarts remained in power, there he was compelled to remain.

Newton had his own troubles with his king. When James took the throne after his brother's death in 1685, he began an inept effort to re-Catholicize Protestant England. In 1687, James took aim at Cambridge University, ordering it to grant Father Alban Francis, a Benedictine monk, the degree of master of arts—an honor that would permit Francis to take an official position in the governance of the university. The university's leaders refused, and Newton applauded. He broke into the last weeks of work on the *Principia* to argue that a "mixture of Papist & Protestants in ye same University can neither subsist happily nor long together." When King James's Court of the Ecclesiastical Commission ordered the university to send representatives to account for its disobedience to the Crown, Newton was selected as a member of the delegation.

The court threatened and blustered. Newton led his colleagues as they pushed back. The government flinched first. In May 1687 the chief judge of the commission issued his order: the Cambridge delegation should "Go your way, and sin no more." Where it counted, Newton and his colleagues had won: Cambridge never granted the required degree.

This victory made Newton a marked man, at least as far as King James was concerned. He returned to Cambridge and, prudently, kept mostly to himself. The fame that the *Principia* brought him was sweet, but for the moment it remained too dangerous to attempt to savor much of celebrity's rewards.

King James II was a failure at most of the arts of governance. He was, however, a master at enraging his enemies and estranging his friends. It took him just three years on the throne to alienate a critical mass of his subjects. By mid-1688, the traditionally pro-monarchy Tories and their opponents, the Whigs, were both conspiring to replace James with his nephew and son-in-law William, Prince of Orange, whose wife was the King's elder daughter, Mary. In November, William landed on the south coast of England with an army of between eighteen and twenty thousand men (including about two hundred black soldiers recruited—or acquired—from plantations in the American colonies). James was able to counter with a force of about the same size and gathered his army at Salisbury, blocking William's path to London, but the royalist strength drained away as first James's generals and then his own daughter Anne defected to William's side. After a couple of minor skirmishes, James ran. He fled London on December 9 and, a week later, surrendered to a Dutch detachment. Two weeks later, William turned a blind eye as his father-in-law escaped to France.

To give his seizure of power its necessary veneer of legitimacy, William summoned a Convention Parliament to settle the question of the royal succession. Cambridge University had two representatives at the assembly. One of them was that newly declared anti-Catholic Isaac Newton.

It cannot be said that Newton was much of a parliamentarian. There is no record of any speech he might have made in the Convention Parliament; his only documented statement on any matter during his year in the House of Commons was a request to a servant to close a window against a draft. No matter, he did what his constituency expected of him, voting with the majority on February 5, 1689, to declare the throne of England vacant by virtue of James's abandoning it, and to offer the unoccupied monarchy jointly to William and Mary.

With that, Newton found himself free to enjoy something genuinely new in his experience: being lionized by the good and the great. He accepted homage from the members of the Royal Society. Christiaan Huygens arranged to meet him and introduced him to the exalted circles at Hampton Court, where Huygens's brother was part of William's retinue. Locke's friend the Earl of Pembroke welcomed him into his home. Newton dined and drank in company that lauded him as the wisest of men and a member of the winning side in what its victors were already calling the Glorious Revolution.

Newton first encountered John Locke as one of those admirers toward the end of 1689, but the two men swiftly formed a bond of deep affection that lasted, with one significant break, until Locke's death in 1704. In most ways, the two men could hardly have been less similar. The reclusive Newton made few friends, and he was a prude—he once dismissed a companion from his acquaintance for telling a lewd joke about a nun. In contrast, Locke played politics at the highest level, lived in the houses of the rich, enjoyed conversation, and took pleasure in the company of women. He was an amiable flirt among wives of repute, addressing one of his great passions, Lady Damaris Masham, as his "Governess."

Nonetheless, the two men did have some connections to each other, notably through Robert Boyle, the pioneering chemist and unofficial leader of London's philosophical circles. Newton knew Boyle as a professional colleague, one of the few he genuinely admired. Locke's connection was more intimate: in the 1660s, still in his twenties and newly qualified as a medical doctor, Locke found in Boyle a kind of intellectual patron and adviser.

The links spread from there. For several years, Boyle had employed as his assistant another young man, the poor but brilliant Robert Hooke. With Boyle's help, Hooke made his way into the center of English science. The Royal Society, founded in 1660, was initially merely a talking shop, in desperate need (so at least some members believed) of someone who would actually do some practical research. In 1662, with Boyle's support, Hooke became the society's first curator of experiments, charged with offering demonstrations three or four times a week. The next year, the society added to Hooke's duties, asking him to keep a daily record of London's weather. Hooke responded with a characteristically effervescent burst of creation, inventing or improving the basic suite of weather instruments: the thermometer, the barometer, rain and wind gauges, and other, more specialized devices. With those instruments in hand, he began to keep his own weather record. Then the thought occurred to him: how glorious it would be if gentlemen of England rose from their beds and made similar observations all over the country, building a picture not just of local conditions but of the varieties of climate throughout the realm.

Hooke published his meteorological call to arms in the journal of the Royal Society, emphasizing the need for rigor: data had to be taken at the same time every day, using instruments whose properties were known and carefully recorded. Robert Boyle thought this a brilliant idea, and he advised his young friend John Locke to enlist in Hooke's crusade.

Locke signed on, devotedly measuring wind speeds, checking temperatures, gauging cloud cover. Doing so, he became, in effect, a foot soldier in what he and his contemporaries understood to be a radically new approach to knowledge. We now call this transformation the scientific revolution, and it is often imagined as a series of heroic battles, victories in a war against ignorance led by men whose names resound like those of triumphant generals—Copernicus, Kepler, Galileo, Descartes, and Newton, the greatest of them all.

But in fact, the shift in understanding that such men led was carried forward through the daily actions of hundreds, then thousands of people who for pleasure, profit, or both set out to use reason and experimentation to order their surroundings. Practical rationalists such as Jethro Tull and his disciples tried to bring the methods of the new natural philosophy to bear on the farm. Amateur naturalists catalogued the habits of animals painstakingly observed over days, weeks, months. One of the more famous among them was Erasmus Darwin; born four years after Newton's death, he absorbed the Newtonian credo that material events must have discernible material causes, and he grappled with the question of the origin of species that his grandson Charles would solve a century later.

England's sailors measured tides, and traders upholding the power of the Crown across the oceans learned mathematics and developed precision tools to measure the motions of the stars and planets. Instrument makers began to establish the crucial idea of standards, common measures that would enable observers anywhere to trust one another's results. Thomas Tompion, the maker of Locke's thermometer, was the first craftsman known to have used serial numbers to identify his finished pieces—bringing science's tools into the nuts and bolts of efforts to systematize the material world.

This was revolution at the barricades: a headlong charge by its partisans to organize, abstract, and universalize their experience of daily life so that its distilled essence would be accessible to anyone who sought it out. Locke, who documented the details of his precision instruments and checked the amount of rainfall and the barometric pressure each day, noting the time of every measurement, was one more cadre in this growing revolutionary band, adding his tiny increment to the arsenal of knowledge.

In the eventful 1660s, Locke had to abandon his first weather diary within a few months. His political career and his own intellectual work consumed all his time and thought. But the experience stuck with him, and more than three decades later, when he retreated from public life for a time to Lady Masham's house in the Essex countryside, he resumed the habits of his youth. It took him some months to unpack his instruments and set up his weather observatory. At last, on December 9, 1691, he made his first observations. Four days later, his weather check had already become routine, a matter of a few minutes each morning.

It had been two years since he had met the unquestioned leader of the new ways of understanding nature, and while Locke had certainly offered explicit homage to his new friend Newton, his resumed weather diary can be seen as a less obvious compliment to the ways of thinking Newton had championed.

Newton's reasons for returning Locke's sentiments were perhaps more simple. Anyone would take kindly to unstinting praise from an intelligent source—and Locke famously evoked affection. When he and Newton finally met, his warmth had its usual effect. Newton's letters to Locke show the impact of Locke's charm: "how extremely glad I was to hear from you," he writes in one; in another, he values Locke's judgment sufficiently to seek his reaction to what Newton called his "mystical fancies"; once he simply admits of "my desire to see you here where you shall be as welcome as I can make you."

In part, he relished the opportunity to tutor so well regarded a man. He gave Locke a private, annotated edition of the *Principia* and composed for him a simplified version of the proof that gravity makes the planets travel elliptical orbits. But Newton's intimacy with Locke seems to have extended well beyond such benevolent displays of mastery. From the beginning, Newton allowed himself to write openly about secret matters. Both men had subterranean interests—in alchemy, for one, the ancient study of processes of change in nature; and in questions of biblical interpretation and belief, which brought them to the edge of what the established English church would damn as heresy.

Locke responded with equal eagerness and candor. He always emphasized his deference on matters of natural philosophy to the man who wrote "his never enough to be admired book." But for the rest, he took part in what became an extended conversation with an intellectual companion, a partner in the pursuit of knowledge of the true nature of the Trinity, about the history of Scripture, about the transformation of substances. And along with his praise and their intense private exchanges, Locke had one thing more to offer: the use of his considerable influence with the Crown.

In the wake of the Glorious Revolution, Locke had become a supremely good man to know. King William cherished him, and he was known and connected by bonds of party and friendship to dozens of the newly ruling elite. He turned down most offers of patronage for himself, but he was perfectly placed to do kindnesses for those he valued.

Newton's service in the Convention Parliament ended on January 27, 1690. He returned to Trinity College and got back to what had once been a satisfactory round of daily life. He worked on corrections to a possible second edition of the *Principia*. He continued to examine the implications of the laws of motion, and he returned to studies of optics and light that had lain fallow for more than a decade. He began to think deeply about the theological consequences of his science, trying to define what kind of God could occupy the universe implied by the *Principia*. It seemed as if he was as much in his natural habitat as ever, wandering through his rooms and his garden, stopping suddenly, when a thought came, to "run up the stairs, like another Archimedes." To outward appearances, this was the man Trinity had sent to London, one who "aim'd at something beyond the Reach of humane Art & Industry."

But the Newton who returned to Cambridge in 1690 was not the same as the one who had set out for the House of Commons the year before. He was not bored, given his impressive productivity over the next few years. But he was restless, unsettled. Cambridge had become small. Its company was dull, uncomprehending of the man in their midst. Notoriously, an anonymous student who passed him on the street said, "There goes the man that writt a book that neither he nor any body else understands." In the face of such indifference (not even disdain!), London's attractions now included company that recognized Newton's worth at something like the value he had come to place on himself. Within months of his return to Cambridge, he let his new friends know he was ready for an escape. There was just one problem: in Cambridge Newton had no material wants. In London he would need to make a living—a good one. How?

Locke knew what to do. Beginning in 1690, he canvassed his most powerful acquaintances to advance his friend's cause. Newton knew what Locke was attempting. In October 1690 he wrote to thank Locke for his efforts; in November he betrayed a hint of urgency, even desperation: "Pray present my most humble service & thanks to my Lord and Lady Monmoth for their so kind remembrance of me. For their favour is such that I can never sufficiently acknowledge it." Such courtesy did not help matters this time—whatever Locke discussed with Monmouth never materialized. But the campaign was under way, with Newton's blessing and ever more urgent hopes.

And so Newton, by candlelight on that cold gray day in December 1691, pushed to one side his angry draft. He took another sheet to try again. "I thank you," he wrote, "for putting me in mind of Charterhouse." He dismissed the idea, but gently this time: "I see nothing in it worth making a bustle for." He summoned the deference due a man in a position to do him good. He begged John Locke to accept "my most humble service & hearty thanks . . . for so frankly offering ye assistance of your friends if there should be occasion."

Days later, when Locke hurried back inside after recording his observations on the weather, careful not to risk his weak lungs and generally frail health on a raw December morning any longer than necessary, it was not Newton's wrath that greeted him. Instead, he read contrite thanks for help given and help to come. Locke took no insult from the rejection of his first attempt, and the letters to and fro confirm that while Newton would remain in Cambridge for five more years, his imagination had already carried him down the road to London. The rest was mere logistics for friends to arrange, to permit the incomparable Mr. Newton to take his rightful place in the big city.

Part II

——◦∞◦——

A Rogue's Progress

5

"The Greatest Stock of Impudence"

WILLIAM CHALONER'S PASSAGE to London came much easier than Newton's. When he decided to go, he walked.

At the same time, his development had some parallels with Newton's. His distinctive qualities of mind made themselves apparent early, in a precocious display of malicious cleverness. Still, as with any great talent, it took years of thought, risk, and practice for Chaloner to achieve all the artful wickedness of which he was capable—an education that he, unlike Newton, had to undertake almost entirely on his own.

Only Chaloner's clash with Newton brought him into history, and most of the details of his early life did not make it into the picture, not even the date of his birth. But the man clever enough to challenge Newton evoked just enough wonder to inspire a sensational biography, written immediately after his execution. Like most true-crime tales then and since, it has to be read with care, as it alternates between admiring horror and respectable condemnation. But at least its anonymous author collected the bare facts of Chaloner's childhood.

He was at least a decade and as much as a generation younger than Newton. He most likely married in 1684, which pushes his birth date back to the 1650s at the earliest, and perhaps as late as the mid-1660s. Like Newton, he was born in the provinces, but his father was poor, a weaver in Warwickshire, in England's Midlands. He had at least one brother and one sister, both of whom he later brought with him into what became a family coining business.

He had had no formal education to speak of, but his biographer noted, "In his Infancy he shew'd a certain aptness to what he afterwards became perfect in." Unfortunately, "as soon as he was able to put any thing in Action, it was some unlucky Rogues Trick or other." At some point, his father and, presumably, his never-mentioned mother found themselves "unable to govern him." They sent him to Birmingham, then a small market town but already known for its metalwork shops and its sketchy regard for the reach of law, to be apprenticed to a nail maker.

Given the apparent trend of his character, they could not have made a more unfortunate choice of trade. Nail-making was at that moment caught between its history and the kind of transformation Adam Smith would make famous a century later in his description of the making of a pin. In Chaloner's day, each nail was still finished by hand, one at a time. The nailer would heat the end of a metal rod in a forge, then hammer the softened tip into a four-sided point. Next, reheating the rod to soften it, he would cut a nail length off. Finally, he would strike the blunt end of the piece to form the head, holding the nail on an anvil or in a tool called a nail header.

All this used to be part of general blacksmithing. But by the time Chaloner entered the trade, nail-making was well on its way to becoming less skilled and worse-paid piecework. The long iron rods were made with a machine called a slitting mill, which was invented in Liège, Belgium, in 1565 and made its way to England around the turn of the seventeenth century. Water power turned two sets of rollers. The first, smooth pair pressed heated bars of iron into thick plates; a second, grooved pair of rollers cut the plates into rods. Those with capital enough to run a slitting mill would advance nail rods to men too poor to pay for them outright, who would then cut an agreed number of nails from a given weight of metal and return them to the mills for a meager payment. Unsurprisingly, those at the bottom of the production line—men who had fire, tools, and a mastery of the basics of working with metal—looked for other opportunities.

Groats, worth four pence, were always rare coins, produced only sporadically by the Royal Mint. A small number were struck in 1561, and later, expanded production from Welsh silver mines led to another issue of the little silver pieces in 1639—these decorated with the ostrich plumes of the Prince of Wales. They were made again from time to time, but few of the coins that were called groats ever saw the inside of the Royal Mint. Instead, private enterprise stepped up, supplying counterfeits—with a notable proportion of the dud money of any denomination produced by men grown weary of turning out twelve hundred nails from every four pounds of iron. Such counterfeits were called Birmingham groats, testimony to the enthusiasm with which the city's metalworkers embraced the craft.

Chaloner's new master seems to have produced his share. Young Will proved a quick study, and soon grasped the "rudiments of Coyning." His teacher did not, however, reap the benefits of his tutelage for long. The son whose father could not govern him was already too ambitious to serve any other man. No later than the early 1680s, William Chaloner abandoned his master and set out on "St. Francis's Mule"—that is, on foot—"with a purpose to visit London." The capital was for him more of a goal than a specific destination. He had no plan, no idea of what to do once he got there.

But the decision to escape to London set in motion the critical phase of Chaloner's education. It would take him the better part of ten more years to master the lessons the city could teach him—the course of instruction that would turn a clever village boy with an elastic moral sense into the man who could present Isaac Newton with formidable opposition.

On arrival, though, even so knowing a young ruffian as William Chaloner would have had no preparation for the shock of London. The city was vast, unimaginably larger than any other place inhabited by English men and women. Its population of almost 600,000, more than ten percent of the national total, was greater than that of the next sixty largest so-called cities and towns combined. Norwich, in second place, was home to between 20,000 and 30,000; at most 10,000 lived in Chaloner's Birmingham.

Seventeenth-century London was a mob of strangers. Its death rate exceeded the birth rate by several thousand a year into the eighteenth century. Yet still it grew, cannibalizing the countryside—drawing from two to three hundred young men and women a day from their villages and towns, come to chase their fortune in the one true metropolis in all of England.

Even the wiliest and most ambitious of these country folk were stunned by their first impression of the capital, which was commonly described as a kind of hell, a "region of dirt, stink and noise." Chaloner would have known he was getting close when he passed the heaps of human and animal waste carted just outside the city every day and dumped along the roadways. Travelers gasped, covered their faces, sped by as fast as they could, gagging.

The city proper brought its own terrors. Prudent Londoners did not drink plain water, especially not from the Thames, for reasons Jonathan Swift made clear in his verses on a rain shower in 1710: "Sweepings from Butchers Stalls, Dung, Guts, and Blood, / Drown'd Puppies, stinking Sprats, all drench'd in Mud / Dead Cats and Turnip Tops / come tumbling down Flood."

Still, while one could live on beer and gin, everyone had to breathe the air. With more than half a million people crammed together, stepping over piles of droppings left by horses, burning wood and coal for warmth, and furnaces, kilns, and ovens making what the city demanded—beer and bread, soap, glass, lime and dyes, pottery, ironwork, and on and on—the atmosphere in the capital was toxic. The resulting "impure and thick mist," if not quite so chokingly fatal as the evil fogs of Victorian London, was still foul enough to drive King William to the suburb of Kensington in 1698.

London did have its rewards, of course: the hope of wealth, or at least of better than subsistence living. The city formed the unquestioned economic center of the nation at a time of radical transformation. It was a fabulously lucrative change: in the late seventeenth century, England fostered a world-spanning web of commerce, with London as its hub. City-based cartels and joint-stock companies pursued their profits in the Baltic and the eastern Mediterranean. Trade with North America was growing. The East India Company had begun capturing all India for the British Crown. Africa, the West Indies, the American colonies, and the home country formed a network that spun slaves, gold, sugar, rum, and cloth around the Atlantic Ocean. The China trade consumed British silver—the preferred precious metal of the Chinese—in exchange for silk and fine ceramics. Almost all of this, three-quarters of England's international trade, passed through London's docks, warehouses, banks, and exchanges.

London dominated the domestic economy as well. Even in years of good harvests, wages in the capital beat those for rural labor by as much as fifty percent. By virtue of its population and its wealth, London formed by far the largest single market in England for food, fuel, cloth, and manufactured goods. Londoners ate sheep from Gloucestershire, drank beer brewed with east-country barley, pulled its herring from the North Sea and cooked them over Newcastle's coal. Interconnected webs of carts, rental horses, and stagecoaches sprang up to carry everything, and London's streets became a tangled mass of animals and people, crowds upon herds, a swaying, shouting, shitting din—exhilarating, terrifying, incomprehensible to anyone encountering it for the first time.

This European urban experience, lived on such a scale only in London and perhaps Paris, formed a network not simply of goods and people but information, from the proper coffeehouse to patronize (Dick's or Will's for Whigs, the Devil tavern or Sam's for Tories), the state of the Baltic market for naval stores, and the most sophisticated houses of prostitution (Mother Wisebourne's establishment off the Strand was a connoisseur's favorite) to the lodes of data that mattered ever more as the city molded the world around it, like the soundings provided by merchant sailors from ports around the world that enabled Isaac Newton to analyze the moon's influence on the tides in the *Principia*. Thus, despite the stench, the sickly living, the fact that there was no place in England where it was worse to be poor, they kept coming, the country-born who overflowed the city's tenements. London's centripetal force, its gravity, was irresistible, and increasing. It was where the action was.

Chaloner's first weeks and months in the big city were typical for newcomers: bad and worse. His biographer reported that on his arrival he found himself "something at a loss of Acquaintance, and knew not what course to take for a Livelyhood." He faced the hard truth that London lived and traded through an intricate weave of associations that seemed impenetrable. Obviously, court or government patronage—Newton's approach—was beyond the reach of a masterless apprentice, and the nexus of trade and high finance even more so. The crafts were also off limits. Though the guild system was weakening in the late seventeenth century, tight networks of skilled men locked out even capable strangers, much less half-trained runaways. As late as 1742, London hatters beat to death a man who dared shape headgear without having gone through the apprentice system. About twenty-five men controlled the cheese trade between London and the major producing region of Cheshire, forcing the hundreds of smaller cheesemongers to accept whatever price the cartel set. The scientific revolution and the incipient industrial one supported a range of new enterprises—precision instrument makers, for one. Chaloner had real dexterity with metal and some knowledge of tools, but even were he willing to submit to a new master, he lacked the bona fides that would have persuaded an established shop to take him on. And so it went, for him as for any unknown newcomer. Despite the high wages available to some, the great mass of London's immigrants found themselves underemployed, battling each other for minute advantages in the daily struggle for existence.

Access to the underworld was tightly controlled too. London's criminals organized themselves into ladders of rank and status as rigid as those of the straight world on which they preyed. Highwaymen like Dick Turpin, celebrated as a kind of latter-day Robin Hood, were the aristocrats. They were generally from higher-class origins than their fellow felons, as they had to have learned how to ride. The tally of those hanged for highway robbery includes parsons' sons, impoverished scholars, spendthrift younger scions of respectable houses—gentlemen, broke, bored, or both.

But if genteel crime was beyond Chaloner, what about the journeyman variety? London in the 1680s and 1690s had no shortage of tempting targets: the cheek-by-jowl jostling between the rich and the ever-renewing crowds of scrabbling poor offered plenty of scope for a little income redistribution. But while the London criminal world was not as organized as it would become early in the next century, it still arrayed itself according to an order a stranger could not simply enter at will. Street thugs evolved specialized techniques for the crush of London's roads. One gang of footpads, led by the marvelously named Obadiah Lemon, taught themselves to use fishing lines and hooks to snag hats and scarves out of moving vehicles. Others preyed on coaches when they slowed at bridges or other obstacles. Pickpockets often started young, practicing on dummies under the eyes of older relatives or friends already skilled in the art, aspiring to rise to the status of "masters of the trade . . . versing upon all men with kind courtesies and fair words, and yet being so warily watchful." They worked in teams, with a carefully worked-out division of labor. One or more stalls (decoys) would lure a cony or a cully (the victim) into a position to be robbed by a foin or a nip. The foins were the elite practitioners, who prided themselves on their dexterity and their ability to distract their victims while they reached into a pocket; the lower-status nips merely slashed and grabbed. Either way, the stolen purse went to the snap, who usually lurked behind or next to the foin or nip, who could then melt into the background.

Shoplifters had a similar battery of tasks divided among gang members. The mask distracted the shopkeeper while the lift grabbed the goods and passed them on to a receiver, or santar, who never entered the store and therefore could not, in theory, be tied to the robbery. Confidence tricks, loaded dice, rigged card games, and the like required similar networks of conspirators. Burglars learned from confederates how to pick locks. Fences, at the hub of criminal commerce, provided a clearinghouse for training, job leads, refuges, and alibis.

In this society of crime, a lone man or woman, unskilled, without friends, known to none of its gentry, would have found it almost intolerably dangerous to attempt a freelance rampage. Chaloner was too smart to try. Instead, he drifted to the hungry fringes of city life until he could find a way to its gilded center.

It took him only a few months, and the path he found earned him scandalized admiration from his biographer, who wrote, "The first part of his Ingenuity showed it self in making Tin Watches, with D—does &c in 'em." These Chaloner "hawk'd about the Streets, and therby pick'd up a few loose Pence, and looser associates."

That is, Chaloner's first attempt to rise above mere subsistence turned him into a purveyor of sex toys. London in the 1690s was as famous, or perhaps notorious, for its spirit of sexual innovation as Berlin would be in the 1920s. Prostitution was ubiquitous, as much a part of the life of the wealthy as it was that of the poor, who supplied most of the trade's workers. The best brothels vied to outdo each other in their range of offerings—so much so that Dr. John Arbuthnot, a man about town in the early eighteenth century, apparently spoke for many when he told a madam at one of the better houses, "a little of your plain fucking for me if you please!"

Anything a cultivated lecher might covet could be had: erotica in words and pictures, ribald songs, and lewd performances. Perhaps the most obscene work of theater ever composed comes from this period, the scabrous play *Sodom, or the Quintessence of Debauchery*, attributed to the notorious libertine John Wilmot, 2nd Earl of Rochester. Written in or around 1672, the play may be a disguised attack on Charles II (with whom Wilmot shared at least one mistress). Its description of a monarch attempting to promote sodomy throughout his kingdom has been interpreted as a coded denunciation of the Declaration of Indulgence of 1672, which pronounced official toleration of Catholicism. If that was the author's intention, the polemic comes very well disguised within its wildly ribald plot.

For those whom literary debauches did not satisfy, a market for sexual aids flourished. As early as 1660, just two years after Cromwell's death, which resulted in the decline of Puritanism, there were reports of imported Italian dildos being sold on St. James's Street. Homegrown entrepreneurs also sought to profit, although it remains something of a mystery just what Chaloner was trying to peddle. That his devices demonstrated "the first part of his ingenuity" suggests that they were more than mere knockoff phalluses. They probably were not true watches either. The technology of watchmaking had advanced a good deal by the mid-1670s. The spiral balance spring, invented by Robert Hooke, stored enough energy and released it precisely enough to allow small, hand-held clocks and watches to keep time accurate to minutes instead of hours—a key step in the evolution of timekeeping. Apprentices usually spent seven years learning the intricacies of clockwork. Balance springs would be used to drive clockwork puppet shows by the early eighteenth century, and it is possible to imagine early attempts to make pornographic displays. Yet it is doubtful that a former nailer's assistant could have mastered the art swiftly enough to begin making his own mechanical automata so soon.

More likely, Chaloner created his own variant of what were then being sold as toy watches. A watch was a mark of status, craved even (or especially) by those who could not afford the real thing. To meet that demand, London craftsmen began making imitations. Surviving examples—most recovered from the tidal margins of the Thames—all have the same basic design: two pieces of pewter, each cast into roughly the shape of half a pocket watch. One half would have a crude dial pressed into it, the other would be decorated with an echo of a gentleman's watch case. The two halves were soldered together and sold as a kind of affordable fashion accessory. Chaloner's metalworking knowledge would have been sufficient for this kind of work—and for the innovation of incorporating dildos into the pieces. He does not seem to have made much money at the scheme. But as his biographer hinted, this brief foray into the fringes of the sex trade was significant not so much for the loose change he managed to pick up as for the looser associates who thereby found him. Some of those new friends aided Chaloner in his next and more successful enterprise, which was built on another basic fact of seventeenth-century urban life: the relentless pressure of infectious disease.

The plague had not returned after the epidemic that ended in 1667, but thanks to London's crowding, its choking air, and its primitive hygiene, deadly disease was always present. Smallpox remained a scourge for both highborn and low. Londoners died of typhus as well—picked up so easily in the prisons that it became known as jail fever. Winters brought tuberculosis and influenza, and in summer, mosquitoes distributed malaria while the swarming flies spread dysentery, infant diarrhea, and more. Children were unbelievably vulnerable. Thirty-five to forty of every hundred children in London died before the age of two. Prosperity was not much protection. The Quakers, a reasonably well-off group and one of the least affected by the plague of cheap gin available throughout the city, lost about two-thirds of their children before the age of five. Virtually all parents would bury at least one infant.

William Chaloner knew a gold mine when he saw it. True medical expertise was expensive, scarce, and often ineffective, while the terror of disease supported a legion of folk doctors, fakers, patent medicine peddlers, confidence men and women. According to his biographer, in order to "satisfy an itching desire he had to be Extravagant," Chaloner found "a Companion little better than himself who had agreed together to set up for Piss-Pot Prophets, or Quack-Doctors."

The key to a quack's success lay in his ability to convince the desperate, and here the young man from Birmingham demonstrated the gift that would advance all his later endeavors. His biographer moralized, but recognized his skill nonetheless: "Chaloner having the greatest Stock of Impudence, and the best knack at Tongue-pudding, (the most necessary Ingredients in such a Composition) 'twas resolv'd that he sho'd personate the Master Doctor, and his Comrade bear the Character of his Servant."

Chaloner starred in the role, able to cajole, to wheedle, to command audiences to accept his authority as a man of uncommon skill and wisdom. His "servant" must have been equally as efficient in coaxing cash out the marks' hands, for Chaloner was soon able to rent a house on the proceeds. He married and sired several children (though it is not known how many, if any, survived). As the years passed, he expanded his repertoire from quack medical advice to a kind of divination, "pretending to tell sill' Wenches what sort of Husbands they should have, discovering Stol'n Goods &c."

This last proved his undoing. There was an obvious trick to recovering stolen property: steal it yourself in the first place. But it took a skilled and careful operator to pull off the scam repeatedly. A few years later, Jonathan Wild would establish control of the London underworld by organizing both sides of criminal undertakings on a citywide scale. He avoided direct participation in the robberies he stage-managed, profiting instead from his roles as restorer of lost property and "thief-taker"—betraying those who robbed out of turn, who competed with him, or who merely had begun to represent a risk to his liberty.

Wild managed this balancing act well enough to dominate the interface of respectable London and its underside for fifteen years. Chaloner, less cautious, blundered. Around 1690, his name surfaced as a direct suspect in a theft. He ran, ending up in the slums of Hatton Garden, anonymous and broke, with nothing better than "some Old Garret to repose his Carcase."

6

"Every Thing Seem'd to
Favour His Undertakings"

DESPERATION NOW DROVE William Chaloner to his last apprenticeship. In his Hatton Garden tenement he met a japanner. The term first described those who varnished or finished surfaces, in imitation of the fine Japanese lacquer work that had reached Europe in increasing quantities over the previous century. From that original sense, it had broadened to include any refinishing involving a hard or opaque coating. Chaloner's neighbor specialized in blacking old clothes with a coating that could restore a degree of respectability to them—if you did not examine them too closely. Selling old clothes to the wretched was a poor man's trade, but Chaloner paid the man to teach him his craft. Then, as Isaac Newton himself would later note in the first entry in his dossier, Chaloner turned himself into a trader in "cloaths thredbare ragged + daubed with colours."

Newton would also later snipe that if Chaloner had remained content in this modest station, he could have avoided his later troubles. But Chaloner had never tolerated mere subsistence, and he had undertaken his new training with a specific end in view. Gilding—the art of coating surfaces with a thin, uniform coat—was a skill that could be applied to more than leather or cloth. In fact, the trade as a kind of archetype of deceit had deep roots. Almost a century earlier, in *The Winter's Tale*, Shakespeare's delusional, jealous Leontes gave voice to suspicions that his beloved son is a bastard, even though the boy looks like him—or rather: "women say so, / That will say anything. But were they false / As o'er-dy'd blacks."

Painting clothes was hard work for little profit. But metal? That was where money could be made. Although there is no record that Chaloner planned out his counterfeiting career, the sequence of his actions strongly suggests that he had figured out the opportunity, probably before his overeagerness as a stolen-goods man got him into trouble. Certainly, he took to his new enterprise quickly, applying his newly acquired technique to silver pieces with which "he thought it Probable to Counterfeit Guinea's, Pistoles [French currency], &c, which being Gilded well [with gold] and Edg'd, might pass for Current throughout the Kingdom."

Adding to the sense that this was a planned rather than an opportunistic move was Chaloner's timing. He saw his chance at the very moment that England was, literally, running out of money in what was a nationwide demonstration of Gresham's law—the axiom that bad money drives out good. The crisis was driven by a peculiarity of England's coinage, the fact that for almost three decades the country had two types of money in circulation: the hand-struck coinage produced up to 1662, and the coins manufactured since on machines installed in the Mint that year.

The older currency, struck by a Mint moneyer swinging his hammer, was irregular and prone to wear. Worse, it had smooth rims, which meant that anyone with a good pair of shears and a file could snip the edge of a coin and then file the piece smooth again. A cut here and a slice there and pretty soon a coin clipper could accumulate a healthy pile of silver, at the expense of a debased currency.

Clipping became epidemic in the 1690s, to the point that at the height of the crisis, "it was mere chance whether what was called a shilling was really ten pence, sixpence, or a groat," according to the Victorian historian Lord Macaulay. In one test of the state of the currency, Macaulay reported, "Three eminent goldsmiths were invited to send in a hundred pounds each in current silver to be tried by the balance." That much money, he wrote, "ought to have weighed about twelve hundred ounces. The actual weight proved to be six hundred and twenty four ounces." And so it went throughout the kingdom: money that should have tipped the scale at 400 ounces actually weighed 240 ounces at Bristol, 203 at Cambridge, and, at rival Oxford, a mere 103.

Clipped coinage was hardly new: it had been punished as high treason since Elizabeth's reign. Ever since, clippers were regularly caught, tried, and condemned to death by the rope or by fire, but with little effect, especially in the orgy of clipping that took place between 1690 and 1696. As Macaulay wrote, the news that one condemned clipper was able to offer six thousand pounds for a pardon "did much to counteract the effect which the spectacle of his death was designed to produce."

There was a still faster route to wealth available to those with access to more sophisticated tools. By 1695 counterfeit money accounted for about ten percent by value of all coins in circulation. The secret of that success lay in the counterfeiter's ability to crack the second and more formidable of England's two types of legal currency.

London had seen nothing like the new moneymaking machinery the Mint installed in 1662. The ubiquitous Samuel Pepys, then secretary to the Navy Board, wangled a personal tour on May 19, 1663. In the narrow Mint rooms crammed alongside the outer wall of the Tower of London, he saw an extraordinary spectacle of heat, noise, smoke, and men pushed to the point of collapse as they raced to keep up with the pace of the giant machines.

In the first room he toured, Mint workers tended intense charcoal fires beneath iron cauldrons, each large enough to melt up to a third of a ton of silver bullion at a time. Other workers poured the liquid metal into sand molds to make small rectangular ingots. Once the ingots cooled, the machines took over. Men broke the molds and passed the silver blocks through huge, crushing rolling mills, powered by horses driving giant capstans one story below. The thin plates that emerged went to lever-driven punches that pounded out round discs to be flattened by a screw press.

Pepys noted approvingly that the new money was "neater . . . than the old way," displaying an unprecedented level of consistency. By law, each shilling was supposed to contain a precise weight of silver, and this mechanical approach linked the symbolic claim of value on the face of the coin—the marks on the coin that declared "this is a shilling"—to the material promise: "this coin contains 88.8 grains of silver, value, twelve pence."

The next machine down the production line was the key to the Mint's ultimate goals and was "the great secret," wrote Pepys, who wasn't allowed to see it. It was one of the earliest edging machines to be used for a national currency. It worked by rotating each coin blank over a pair of steel plates. As a Mint moneyer turned a hand crank connected to a toothed cog, the plates engraved an inscription into the rim of each coin—on the larger denominations, the phrase that still figures the edge of the British one-pound coin: *decus et tutamen*, "a decoration and a safeguard." This was the Mint's secret weapon, for edged or "milled" currency could not be clipped without the crime revealing itself in a gouge in the milled rim.

The final step in the process—striking the coin faces with the appropriate image—was newly mechanized as well. A Mint worker sitting in a pit below floor level would place a coin blank into the striking chamber. Four men pulled on the ropes at either end of a huge capstan, causing its arms to spin and the press to drive two steel dies into the faces of the coin, producing an impression deeper and sharper than a man with a hammer could hope to match. As the dies retracted, the moneyer sitting in the pit would flip the newly struck coin out of the chamber and replace it with a new blank.

When driven hard, the teams at the presses could produce a coin every two seconds. Even at somewhat slower speeds and with every trick of mechanical advantage brought to bear, however, the machines consumed men. Those at the capstan bars exhausted themselves within fifteen minutes, and losing fingers was reckoned to be an ordinary hazard for those who fed blanks into the striking chambers. Perversely, the brutality was part of the appeal of mechanizing the Mint. If a trained crew found it so draining to produce the new coins, then would-be criminals should find it nearly impossible to copy them. As Pepys approvingly concluded, the new machines yielded a currency that was "freer from clipping or counterfeiting" than ever before. No mere London ne'er-do-well could copy coins "without an engine of the charge and noise that no counterfeit will be at or venture upon."

Pepys drastically underestimated the ingenuity of England's underworld. William Chaloner, for one, was already comfortable working with hot metal. And in a goldsmith named Patrick Coffee he found one last master, who taught him the essential techniques of counterfeiting over a period of several months, probably extending into early 1691. Neither Coffee nor Chaloner left notes describing exactly what was learned in this final apprenticeship, but surviving trial records of coiners prosecuted at London's criminal court, the Old Bailey, in the late seventeenth century provide a clue.

Those cases show that it was as important to learn what not to attempt as what to do; inept counterfeiters attempting to exploit the currency crisis supplied the Old Bailey with a constant diet of rapidly dispatched defendants. Perhaps the most spectacular display of incompetence came from an unnamed "inhabitant of the parish of St. Andrews Holbourn," brought to trial accused of copying French coins. His work was astonishingly awful, and he was acquitted, the jury accepting his rather bold argument that the poor quality of his work confirmed that "he had tryed to Coin with Pewter as afore-said for Diversion, or the like, but never was concerned in Coining any manner of Money." Few others tried this defense.

Mary Corbet was more typical. She faced the court on April 9, 1684, charged with transforming "twelve Pieces of Copper, Tinn, and other false Metals, into the Likeness and Similitude of the currant Coin of this Kingdom, called Queen Elizabeth's Shillings, and twelve other pieces of like counterfeit Metal, into the Likeness, &c. of Queen Elizabeth's Six-Pences." Corbet had taken the prudent course: the pre-1662 hand-hammered money was so debased, as early as 1684, that making facsimiles of the irregular coins presented fewer challenges than trying to mix up a batch of the new, milled coins. Two witnesses testified that Corbet melted "a certain Quantiy of Pewter, Copper, and such like Metals, (as much at a time as weighed about a Pound weight) in an Earthen Porringer," and that she then poured her hot metal "into Moulds of Wood, which fashioned it like the Shillings and Six-Pences before mentioned." That was it: Corbet ran her coining operation with a heatproof pot, a vigorous fire, and a simple mold.

As usual, the production side of the counterfeiting business was the easy part. The dud coins had to be spent, and here a better choice in her confederates might have saved Corbet. The witnesses against her were two women caught trying to pass her products. Based on their testimony, she was found guilty of high treason, for which the penalty—for women—was to be burned alive (though in many cases "mercy" was granted and the condemned would be strangled and her corpse burned). Corbet appealed the sentence, "pleading her belly"—that is, that she was pregnant—in the hopes of receiving a reprieve at least until the notional child was born. A jury of women was impaneled on the spot, felt her abdomen for any signs of a moving fetus, found none, and left her to her fate.

There are dozens of cases like these in the record, a litany of marginally capable small-timers, regularly caught and brought to trial. A surprising number ended in acquittal, often in the teeth of powerful evidence. The very severity of the punishment worked against the authorities at times. Certainly, even as Parliament increased the penalties for the variety of coining crimes—owning coiner's tools, possession of bad coins, and so on—juries would go to considerable lengths to avoid condemning appealing or reputable defendants.

Occasionally, though, much more sophisticated operators would stumble, and the scale of their operations—the amount of damage they could do—usually ensured a concerted effort by the prosecutor to leave no room for misplaced mercy. As a consequence, a few trial records contain fairly detailed accounts of the way expert coiners worked.

For example, there was the case of Samuel and Mary Quested, brought before the Old Bailey on October 14, 1695. They were charged with "forging and coining 20 false Guinea's 100 King Charles the First shillings, and 10 King James's the Second Milled shillings." According to witness depositions, Samuel Quested had practiced as a coiner for several years. At first, following the historical evolution of the Mint's methods, "when he coined money he struck it with a hammer." But producing copies of a rapidly evaporating currency was not nearly as lucrative as forging modern, milled coins, so Quested, probably aided by his wife, created their own version of the official production line. When Mint agents searched the cellar of Quested's house, "they found a Cutter to cut out the Plate fit for Coining"—the machinery needed to complete the first step in the minting process, cutting accurate coin blanks. Next the investigators passed out of doors, where "they found the Dyes to make Guinea's, Shillings and Halfpence, in an old Stocking hid under the earth." These, combined with another discovery—"in an Outhouse they found a Press to coin money with"—showed that the Questeds had the capacity to strike a design in deep relief on the faces of their coins. One more critical process remained—and the Mint men continued their search until "in the Orchard they found Instruments to mill the Edges of Milled money, and divers other Dyes."

With that, Samuel and Mary Quested could complete near-perfect imitations of the King's coins, down to the edging that was a decoration, but no protection. Witnesses deposed that they had seen Mary Quested mill guineas made out of what was described as "a coarse sort of gold"—most likely clippings from valid coins mixed with pewter or brass or other base metal. Whatever the precise metal recipe used in the Quested workshop, their products were clearly of the highest quality. Their false guineas traded for twenty shillings each, which, as the court record grimly noted, was the same value "as guineas go now."

From the start of his coining career, William Chaloner aimed for the kind of perfection the Questeds had achieved—coins that could pass without detection. Patrick Coffee had brought him almost to that point. Based on the evidence of the coins he eventually produced, Coffee's instruction enabled Chaloner to prepare metal plates for punching into coin blanks. Coffee also taught him how to build and use a serviceable coin press, one able to strike a deep engraving on the front and back of each coin; and, at a minimum, he showed him how to use molds to produce a plausible imitation of the milled edging that was the Mint's pride and the clipper's enemy.

With that, Chaloner would have been almost—but not quite—ready to begin coining at the highest level. He still lacked one crucial tool. The quality of his finished coins depended on how well a coiner could reproduce the design of a legitimate coin's face. To do so, a coiner needed near-perfect dies for use in his coin press. It took a master engraver to make such dies, with a level of skill far beyond what either Coffee or Chaloner could muster on his own. Chaloner found his die maker in Grays Inn Lane, at the workshop of the print seller and engraver Thomas Taylor.

At first glance, Taylor seems an unlikely recruit. For the most part, he is remembered as an honorable minor member of the society of learned men—the so-called Republic of Letters—known to his contemporaries for publishing the landmark map collections *England exactly described* and *The Principality of Wales exactly described*—two examples of the new demand at that time for more precise renderings of the physical world. In 1724, as part of a wave of post-Newtonian popularization of astronomy and physics, Taylor produced a broadside illustration of a solar eclipse, complete with diagrams explaining the orbital geometry that lies behind each total eclipse.

Reproducing that kind of detailed imagery required highly specialized skills. But for all his ability, Taylor seems to have found, as many have since, that publishing is a tough business. Selling maps out of taverns (he advertised the Golden Lion on Fleet Street as one of his business addresses) did not cover the bills. So, at least in his youth, Taylor was willing to provide his services to William Chaloner. His workmanship proved first rate. In 1690, Chaloner took delivery of dies for French pistoles —gold coins worth about seventeen English shillings. In 1691, he came back for more and commissioned Taylor to make him a second pair of dies, these engraved with the face and reverse designs of golden English guineas.

At last Chaloner was ready for his first major attempt at currency crime. He produced his initial run of counterfeit coins with his new tools and knowledge sometime in 1691, using an alloy of at least moderately pure silver to strike a large run of pistoles (thousands at least) followed by "a great number of

Guineas, all of silver gilt." He still needed help—Coffee and another man, his brother-in-law Joseph Gravener (who also answered to Grosvenor), performed the final step of gilding the pieces with a thin skim of gold. Chaloner's chief confederate, Thomas Holloway, together with his wife, Elizabeth, handled the crucial task of passing the false coins into the hands of the petty crooks who put the counterfeits into circulation. Holloway later confessed that he had had no trouble unloading the merchandise. The amounts involved were impressive: Holloway said that he had taken "at least 1000 of Chaloner's French Pistols and . . . hath seen many roods of his Guineas," which Chaloner sold for eleven shillings each.

Holloway also said that he had heard Chaloner "boast his workmanship"—and with cause. Chaloner could not keep up with the demand. His biographer reported that, having "thought it Probable to Counterfeit" good money, now he raced to produce as much as he could. "The Trade went briskly," and *"Chaloner's Guinea's* flew about as thick as some years ago did bad Silver." The payoff would have been enough to catapult Chaloner into real wealth almost overnight. Using Holloway's numbers, Chaloner's profit in his first several months would have exceeded one thousand pounds, or roughly twenty times the annual wages a skilled laborer in London could expect. These were high times, when "every thing seem'd to favour his Undertakings," and "he seem'd to have found the (so-much-sought-after) Philosopher's Stone; or (like Danae from Jove) had Showers of Gold daily falling into his lap."

William Chaloner responded to this sudden fortune with glee and appetite, plunging into all the pleasures now open to him. He reveled in the company of women, and as his biographer noted with his characteristic mix of awe, envy, and condescension: "To compleat (as he thought) his happiness, he wanted nothing but a Phillis, (For a Coyner, you must know, is as rarely to be found without a Harlot, as a Sea Captain's Wife without a Gallant)." London had no shortage of willing (or purchasable) candidates, and the open-handed Chaloner soon found the first of a series of mistresses. "As the Devil would have it, a Phillis he found, every way fit for himself." And so "he forsook his Wife, a very good Woman, by whom he had several Children," to take up with a woman whose parents "lovingly Bawded for their Daughter, and she as dutifully ask'd her Mother's Blessing when she went to bed with her Spark."

At first Chaloner occupied himself with his new companion at her parents' house, but the tide of cash soon carried him to grander quarters. As Newton later disdainfully reported, "in a short time [Chaloner] put on ye habit of a Gentleman," and set out to cloak himself in all the outward appearances of respectable wealth. He found a pleasant address, "a great house in Knightsbridge"—then a semirural suburb—which he ornamented with that conventional blazon of sufficient fortune: his own service of (presumably real) silver plate.

Such ostentation was dangerous. Chaloner's boisterous display of new wealth lifted him out of the anonymous ocean of London's poor. He was becoming a name attached to a face, visible well past his immediate circle of henchmen and targets. That made him vulnerable to anyone moved or compelled to inform. Chaloner's first foray as a coiner lasted about two years—a long run for such a high-volume operation. But he came to grief in the usual manner, when William Blackford was caught and tried for passing false guineas. Blackford was condemned to death, but bought himself a reprieve by implicating the man who supplied him with "some hundreds (if not thousands) of his Guinea's and Pistoles": Mr. William Chaloner. After he got word of Blackford's testimony, Chaloner spent two days coining as many guineas as he could, to supply him in the long drought to come. Then he stashed his precious dies and some other equipment with Thomas Holloway and disappeared.

He stayed underground for five months. Blackford remained in his holding cell in prison, ready to testify, but eventually his jailers tired of waiting for the elusive Chaloner and satisfied themselves with the counterfeiter in hand. Blackford was hauled to Tyburn and hanged late in 1692. Shortly afterward, Chaloner resurfaced, but he did not immediately restart his coining production line. After his enforced vacation, he seems to have been short of the requisite capital to pay for the sophisticated tools required, along with the significant amounts of silver and gold needed for first-class work.

Instead, he found a new source of funds in treachery. William III, who had so recently deposed James II, still feared his rival's return. Jacobite sedition—as James's cause was called—sparked major scares and some minor actual threats in London throughout the early 1690s, and the government offered rewards for information about treasonous conspiracies. Chaloner recognized found money when he saw it, and in mid-1693, he set out to find those whom he might usefully betray.

He approached four journeymen printers with copies of King James's declaration of that April, which sought his return to the throne and promising free pardon to his opponents, lower taxes, and liberty of conscience to all his once and future subjects. They refused to print the incendiary sheet. So Chaloner composed a new Jacobite document of his own and "importun'd them to Print him some," promising that this new pamphlet would be distributed only privately to Stuart sympathizers. Two of the printers resisted, so Chaloner turned his attention to the other two—a Mr. Butler and a Mr. Newbold. It took Chaloner "ye expense of several threats and some money," but "at length they were prevail'd on." Chaloner had them deliver the pamphlets to the Blue Posts tavern in Haymarket, and he invited his co-conspirators to join him there for a celebratory dinner. The printers dined—well, one hopes—and then "instead of Grace after Meat, [Chaloner] entertain'd them with Messengers and

Musqueteers and Swearing the Fact at the Old-baily." At trial, Butler and Newbold were convicted of high treason and condemned to death.

For this service Chaloner was promised one thousand pounds from a grateful Crown and government—or, as he later bragged, "He had fun'd (that is tricke'd) the King of 1000." Always happy to milk a willing cow, Chaloner now pursued a career as a professional informer, to the point of voluntarily going to jail for five weeks to eavesdrop on Jacobite prisoners. But he never repeated his initial coup—several of the prosecutions launched with his information did not produce convictions—and his cash flow ebbed.

This game ended for good when he crossed paths with a man almost as unscrupulous as himself, a thief-taker named Coppinger. In the loosely policed city, thief-takers filled a vacuum by serving as bounty hunters, tracking down criminals on their own initiative in return for fees from crime victims and rewards from the state. The potential for abuse was obvious—Chaloner's own brief period as a discoverer of stolen goods illustrated how easy it was to play both sides, organizing crimes and betraying gullible accomplices.

When Chaloner met him, Coppinger had turned to extortion, soliciting bribes and "forcing Money from people, under pretence of Warrants to apprehend 'em." Captured and committed to Newgate, he tried to buy freedom by implicating bigger fish among the coiners he had known through his line of work. According to Coppinger, "the said *Chaloner* being once in Company with him, accosted him after this manner: Coppinger, *I know you have a pretty knack at writing Satyr* [satire]; *do you write something against the Government, and I'l find a Man shall Print it; then you and I'll Discover it, whereby we shall take off all suspicion of being guilty of any Crime to the prejudice of the Kingdom.*" Coppinger reported this to the Lord Mayor himself, and Chaloner landed in a holding cell in Newgate.

Chaloner's famed "knack at Tongue-pudding" served him now. He told tales in his turn of Coppinger's talent as an extortionist. The case against Chaloner never materialized—there was no evidence beyond one man's word against another—and on February 20, 1695, Coppinger was brought before the Old Bailey to face trial. He complained of being "maliciously prosecuted," and asserted that the witnesses against him were known coiners worse than he. He could not, however, explain how a watch worth four pounds that supposedly belonged to one Mary Mottershed had found its way into his possession. He was pronounced guilty of felony theft and sentenced to death. Chaloner walked away untouched.

William Chaloner read his own lesson in his seeming invulnerability. In 1693, shortly after his Jacobite printers were condemned, he started up his counterfeit mint once more. His confidence in his ability to outwit the Mint—certainly as long as it was staffed by the usual mix of patronage absentees and corruptible underlings—had been confirmed. His coining trade flourished, forcing him to add employees to keep up with the work. He trained "Relations, nay almost all his acquaintance to do something relating to it." For the moment, Chaloner reigned in his corner of London, a kind of criminal alchemist, able to multiply without limit coins that looked so persuasively like true silver and gold.

Part III

—◦◦◦—

Passions

7

⎯⎯∞⎯⎯

"All Species of Metals . . . from This Single Root"

THE HONORABLE ROBERT BOYLE, seventh son and fourteenth child of the Earl of Cork, had been ill throughout 1691. In July the situation became grave enough to jolt him into writing a will. By Christmastime it was clear that the great chemist (and notable experimental physicist) was dying.

Boyle's intellectual output had been prodigious. Just as important for the future of British science, he had the gift of recognizing and supporting brilliance when he saw it. Boyle had been Robert Hooke's first patron, John Locke's mentor, the young Newton's occasional correspondent. For three decades and more, he had been the living center of London's learned life. But his decline was no great surprise to those who knew him well. He had been sickly as a child and frail ever after. He had dodged the worst afflictions of the great plague epidemic of the mid-1660s and the more routine ebb and flow of the other infectious diseases that carried off so many contemporaries. But he had endured almost everything else: fevers in and out of season, excruciating recurrent kidney stones, a stroke that left him temporarily paralyzed, although he had continued to dictate experimental procedures to assistants as he recovered.

He was a man of deep and committed Christian faith. He believed in the resurrection and the glory of God and the joys of the world to come. But if death should have held no terrors for him, Boyle was human enough to admit to fear of the pain of dying. He was fortunate in this, as in so much else. Late in the day on December 31, his life ended calmly, with no evident distress, in his bed at the grand house on Pall Mall.

Isaac Newton set out for London the day after Boyle's death, and almost certainly went to Boyle's funeral at St. Martin-in-the-Fields on January 7. Two days later, he dined with fellow mourners, including Samuel Pepys and his fellow diarist John Evelyn—another of the founders of the Royal Society—and their conversation turned to "thinking of a man in England fitt to bee set up after [Boyle]" as the leader of the nation's intellectual life.

The obvious candidate, of course, was sitting at that dinner table. But the right position in London still eluded Newton. Also, unknown to Pepys and Evelyn, the immediate consequence of Boyle's death was to force Newton to confront anew work both he and Boyle had attempted—and kept almost entirely hidden—for two decades.

Death bursts secrets, and this one began to crack just weeks after Boyle's funeral. In February 1692, Newton wrote to John Locke, mostly to announce that he would abandon hopes for a patronage job for a while. But in the last line of what reads like a hurried postscript, he notes that Locke—one of Boyle's oldest friends—had taken possession of something he cryptically called "Mr. Boyles red earth."

Locke's reply has been lost, but apparently he picked up on the hint and sent Newton a sample. Then a ripped and partial letter from Newton in July seems to warn Locke off sensitive ground. He writes that he had received too much of the earth, "For I desired only a specimen, having no inclination to prosecute ye process." But, he added, if Locke wanted to attempt the experiment, then he would try to help, "having a liberty of communication allowed me by Mr. B[oyle], in one case wch reaches you." Newton said that he stood obliged to Boyle to preserve this secret—and presumed that Locke, equally Boyle's confidant, accepted the same obligation. The implication is obvious: the process employing the red earth was incredibly sensitive and could not be discussed unless Locke committed to a vow of silence.

Locke replied swiftly. He assured his friend that he had been initiated into the mystery: Boyle had "left to . . . me the inspection of his papers"—including the ones never intended to reach the public. To reassure Newton, he enclosed copies of "two of them that came to my hand, because I know you desired it." One document survives. It describes in fairly clear language the series of steps through which one might purify the element mercury: wash it repeatedly with a particular soap that would, Boyle wrote, force it to "throw out any feculency that may lie concealed in ☿ [mercury]."

Simple as it sounded, Boyle's experiment clearly fascinated Locke, and Newton felt compelled to give his friend one last warning. To his certain knowledge, Boyle had first examined this process as much as twenty years earlier, "and yet in all this time I cannot find that he has either tried it himself or got it tried with success by anybody els." Newton, for his part, wanted nothing to do with it. He was glad Locke had received instruction from Boyle's papers, for "I do not desire to know what he has communicated but rather that you would keep ye particulars from me . . . because I have no mind to be concerted with this ℞ any further than just to know ye entrance." Locke could go ahead if he wished, regardless of Newton's efforts to "perhaps save your time and expense." Yet for all his studied lack of interest, Newton allowed he had a project of his own: "I intend . . . to try whether I know enough to make a ☿ [mercury] wch will grow hot with ☉ [gold]."

To find a substance, some "mercury," that will interact with gold? Now Newton was getting to the nub of the matter. Boyle's reluctance to part with all he knew, even to Newton; Newton's initial circumspection with Locke; Locke's own withholding of the larger and more provocative part of the process—all these derived from the fact that the three men were talking—or rather, trying not to talk about—one of the deepest mysteries of the natural world. William Chaloner was not the only man in England searching for a way to create wealth without limit. The esoteric recipe hidden in Boyle's papers—or so Newton and Locke hoped, or wondered, or doubted—contained a method by which someone adept in the manipulation of matter and heat could transform base metal into pure, lustrous, immortal gold. In other words: alchemy.

From this distance, after almost three hundred years of systematic chemistry, alchemists seem little better than con men, or at best self-deceivers. To modern eyes, alchemy is groundless superstition, the same sort of unreason that led some of Newton's contemporaries to fear the occult powers of witches.

In fact, alchemists had a bad reputation by Newton's day too. Ben Jonson parodied them as greedy charlatans in *The Alchemist*, first performed in 1610. His hero, Subtle, wields a half-mastered patter of alchemical jargon in order to gull the gullible and win the affections of a comely nineteen-year-old widow. He openly counterfeits: to persuade one reluctant client to hand over the last of his money while he waits the two weeks or so for the alchemical process to produce cartloads of gold, he offers "a trick / to melt the pewter, [that] you shall buy now, instantly / and with a tincture make you as good Dutch dollars as any are in Holland."

And yet, Robert Boyle, who was neither a criminal nor much given to folly, pursued alchemy with passion. So did Isaac Newton for more than twenty years, with all the concentration and effort that he devoted to mathematics or physics, producing more than a million words of notes: queries, copies of older texts, and page after page of laboratory results. He and Boyle and Locke—and dozens of others throughout Europe—still felt the urgent need to mix and shake and heat and cool compound after compound, in pursuit of something more valuable to them than mere gold. Why?

Because, at least for Newton, alchemy offered two prizes of infinite worth. The first was the usual aim of Newton's investigations: knowledge of the created world. Alchemy, as he and Boyle approached it, was an empirical, experimental science. Its theory was occult—literally, hidden—but its practice was hard, hot, and practical, the manipulation of matter with heat, solvents, weights and measures. Each alchemical experiment told Newton some fact about the behavior of the physical world.

That was an end worth seeking on its own terms, but it was the second aim of the work that drove Newton's periodic near-obsessive concentration on alchemy. Newton understood the implications of the expanding reach of natural philosophy, of course—none better. When he first encountered the mechanical worldview, he had concluded that it made no sense to declare that "ye first matter" derived from any prior source, "except God." He crossed out those last two words, it's true—but he had written them down first.

And in them, Newton recognized the essential fact that remains at the core of modern science with its material explanations for physical events. In a world composed entirely of matter in motion, the traditional role of God had to shrink. The author of a mechanical universe could put events in train, but after that primary impulse, the cosmos could then wend its way forward through time on its own.

It was not just Newton who felt the chill of an increasingly Godless nature. Every careful observer understood the implications of the new approach. The year after Newton was born, one of its central proponents, René Descartes, had to defend himself against charges of atheism. In 1643, Martin Schoock, a professor of philosophy at the University of Groningen, in the Netherlands, bitterly condemned Descartes as the "prince of Cretans" (from the old gibe about the man from Crete who assures his hearers that he speaks the truth when he says that all Cretans are liars); for being a "lying biped"; and, worst of all, because "he injects the venom of atheism delicately and secretly into those who, because of their feeble minds, never notice the serpent that hides in the grass."

To Schoock, the sin lay less with Descartes' physics and more with his reverence for the power of human reason. He was particularly suspicious of what he saw as the Frenchman's strangely weak affirmation of the existence of God. (Descartes complained of the unanswerable nature of the charge to the French ambassador to The Hague, writing that "simply because I demonstrated the existence of God, [Schoock] tried to convince people I secretly teach atheism.") Descartes himself escaped serious consequences. But the stench of atheism stuck to the new science—and by the time Newton first came into contact with Descartes' work, the implications of a physics that virtually eliminated the need for God to act in history were obvious even to a youth just starting to read the basic texts on the fringes of the educated world.

Newton ultimately demolished Descartes' physics, and long before that he had found a way, satisfying at least himself, to restore God to the center of the action in space and time—most dramatically, perhaps, in his arguments for why the sun and the planets should experience their mutual gravitational attraction.

His early writings about how divine action shaped the solar system were still a bit vague, as in the letter he wrote in 1675 to Henry Oldenburg, secretary of the Royal Society, in which he suggested "so perhaps may the Sun imbibe this Spirit copiously to conserve his Shining, & keep the Planets from recedeing further from him." But Newton had sharpened his view by the time he came to the *Principia*. Gravity, he argued, derived from divine action. There, he invoked the presence of God directly, declaring that when the tails of comets brushed past the earth, they deposited that spirit "which is the smallest but most subtle and most excellent part of our air, and which is required for the life of all things."

As Newton developed his thinking, his new physics grew ever more hospitable to his vision of an omnipresent, omnipotent, all-knowing, and, above all, an active deity, fully present in the material cosmos of space and time. He explicitly offered the *Principia* as testimony to the existence and glory of all-creating divinity: "When I wrote my treatise upon our System, I had an eye on such Principles as might work wth considering men fore the beleife of a Deity," he wrote to Richard Bentley, an ambitious young clergyman preparing the first of the series of lectures Robert Boyle had endowed in defense of Christian religion. "Nothing can rejoice me more," Newton added, than that his work would prove "useful for that purpose."

Finally, in 1713, Newton expressed his mature conception of divine action in a short essay added to Book Three of the second edition of the *Principia*. Called the "General Scholium," it contains a passionate account of God triumphant in nature. He wrote, "This most beautiful system of the sun, planets and comets could only proceed from the counsel and dominion of an intelligent and powerful Being." How smart? How powerful? "This Being governs all things"—and Newton meant governs—"not as the soul of the world, but as Lord over all." What are his attributes? "The true God is a living, intelligent, and powerful Being . . . He is eternal and infinite, omnipotent and omniscient." Where does this God reside? "He endures forever and is everywhere present . . . He is omnipresent not virtually but also substantially."

This was a God to animate the dry bones of mathematical philosophy. Existing everywhere, for all time, he is "all similar, all eye, all ear, all brain, all arm, all power to perceive, to understand, and to act." All this within a cosmos that Newton elsewhere called his "boundless, uniform Sensorium," within which God could "form and reform the Parts of the Universe."

That is: Newton's God existed everywhere, "substantially"—really, materially there, able to impinge on matter instantly, through all of space and time. The observed fact of cosmic order, combined with Newton's demonstration that human mathematical reason could penetrate that order, implied (necessarily, to Newton) the existence of that perfect being from whom both order and intelligence derived. Newton's natural philosophy was thus, as he had told Bentley, explicitly an inquiry into what could be discovered through the properties of nature about the divine source of all material existence.

Newton was convinced. Nonetheless, some uncharitable louts remained unpersuaded, disdainful. Leibniz, for one, ridiculed the notion of a divine sensorium and what he saw as Newton's flight to an occult explanation for gravity. What was wanted, what Newton sought, was an eyewitness demonstration of divine action in nature.

Hence, alchemy. Alchemy seemed to offer a way for him to rescue his God from the threat of irrelevance—salvation through the ancient alchemical idea of a vital agent or spirit. This vital spirit, Newton wrote, had all the attributes of God. It was omnipresent—"diffused through everything in the earth." It was enormously powerful, destroying and creating throughout nature: "when it is introduced into a mass of substances its first action is to putrefy and confound into chaos; then it proceeds to generation." In the conventional language of alchemy, this cycle of decay and growth was called vegetation. "Nature's actions," Newton wrote, "are either vegetable ... or purely mechanicall." In contrast to mere mechanics, vegetation animated matter, as the vital spirit served as "her fire, her soule, her life."

Pared down to its essence, Newton's quarter of a century of alchemical experimentation formed his attempt to capture the active, vegetative spirit through which God translates divine intention into the shapes and changes of the living world. He annotated his hermetic texts, interpolating thoughts about the process of vegetation, of the vital, living spirits that propelled its changes, and, above all, of God as the first author of living transformation. Then, from his writing desk upstairs in his rooms at Trinity, he would take those secret thoughts to the shed next to the chapel, there to seek tangible proof of that divine, ubiquitous, active presence.

He had kept at it all those years—parts of four decades—because here, he believed, he could actually demonstrate how God continued to work in the world. He said so explicitly in a note from the 1680s: "Just as the world was created from dark Chaos through the bringing forth of the light and through the separation of the aery firmament and of the waters from the earth," Newton wrote, checking off the boxes of the first chapter of Genesis one by one. "So our work brings forth the beginning out of black chaos and its first matter through the separation of the elements and the illumination of matter."

His work! Human hands, his own hands and eyes and brain, to bring beginnings out of impenetrable chaos? No one should ever say that Isaac Newton was passionless: this is the cry of an ecstatic, as extravagant in his dreams of communion as any desert-maddened hermit. But strip away what borders on hubris, too close an imitation of God, and what remains is Newton's essential ambition: to replicate divine action closely enough to provide incontrovertible, material proof of the fact of God's work in Creation and ever after.

He knew that all the theorizing, all the theological argument, all the indirect evidence from the perfect design of the solar system could not match the value of one actual, material demonstration of the divine spirit transforming one metal into another in the here and now. If Newton could discover the method God used to produce gold from base mixtures, then he would know—and not just believe—that the King of Kings would indeed reign triumphant, forever and ever.

8

"Thus You May Multiply to Infinity"

THERE IS A FAMOUS picture of Trinity College made during Newton's tenure there. In the foreground, just before the college gate, two men talk while a couple of dogs wrangle. A few members of the college walk the paths of the Great Court, and near the northwest corner of grounds, someone tends a small bonfire. The architecture looks familiar, much the same as now, but there is one detail that has long since been lost: a crude little structure tucked against the choir end of the college chapel, close beside Newton's rooms. Almost certainly, this cramped, dark shed was where Newton kept his alchemical laboratory.

Newton's first alchemical research began in 1668. He returned to it for extended periods over the next quarter century. He kept the work quiet, fully committed to the alchemical tradition of secrecy. When Robert Boyle announced that he planned to publish some of his results in the Royal Society's *Philosophical Transactions*, Newton was horrified by the breach of security. There were practical reasons for his caution: as Ben Jonson's gibes demonstrated, alchemy looked an awful lot like counterfeiting to the uninitiated. In fact, such experiments were against the law, England's Act Against Multipliers, a statute whose repeal Boyle himself arranged in 1689. But worse, from Newton's point of view, was the notion of exposing potentially divine (and hence extremely powerful) secrets to the vulgar masses. If the process Boyle described had applications beyond heating gold, its description could cause "immense damage to ye world." Newton added—or rather warned—"I question not but that ye great wisdome of ye noble Author will sway him to high silence."

As measured by the time, effort, and accuracy of his laboratory trials, Newton was by far the most sophisticated and systematic alchemist in history. Most other genteel alchemists, even Boyle, relied on assistants to do the messy side of the work. Newton himself performed the tedious sequences of grinding, mixing, pouring, heating, cooling, fermenting, distilling, and all the other manipulations required. He even designed and built with his own hands the furnaces within which his alchemical reactions took place.

Above all, he demanded a level of empirical precision that no other alchemist had ever attempted, and he pursued that experimental rigor with manic, total devotion. Humphrey Newton described what went on there as a continuous, almost industrial operation: "About 6 weeks at Spring and 6 at ye fall, the fire in the Elaboratory scarcely going out either Night or Day, he siting up one Night, as I did another till he had finished his Chymical Experiments." For every experiment, Newton recorded to the grain the amount of each input, and measured its products to the limit of his instruments to resolve. Newton repeated his experiments as often as necessary, Humphrey Newton reported—never mind the heat, the fumes, the choking smoke that alchemical trials routinely produced. Through it all he never violated the adept's code of secrecy, even to his own servant: "What his Aim might be, I was not able to penetrate."

Newton even broke away from writing the *Principia* in 1686 and again in 1687 to keep his regular spring appointment with his fires and crucibles. Looked at one way, for much of the 1680s Newton was most concerned with his urgent study of the mutability of metals, and that little detour into physics and mathematics after Halley's visit in 1684 was merely a distraction, a break in the flow of the true course of his work. He did pause at last, once the fame that followed the *Principia* consumed him.

But then, in 1691, after his return to Cambridge, Boyle died. Some weeks after that, Locke wrote to Newton about that mysterious red earth, and once the samples arrived, the furnaces seized hold of Newton once more.

That summer, he recorded the first in a new series of experiments. Over the next two years, he chased Boyle's process through his laboratory in a sustained frenzy of work, one that would prove to be his last major attempt to coax the vital spirit into revealing the secret of the transformation of base mixtures into gold.

Every document Newton wrote offers a probe of his mind, each one a partial view, a snapshot. Some, like the papers that lead up to the *Principia*, give us the Newton of popular memory. We recognize the brilliant, logical, dispassionate thinker, working more or less systematically, discarding failed, older conceptions as he moved toward a goal that became ever more clearly defined over time.

Newton's alchemical writing offers another view. As ever, he wrote, with his near-graphomane's determination to draft and redraft until he achieved the precise shade of meaning he wanted. One survey of his writings between the completion of the *Principia* and his final departure from Cambridge in 1696 counted at least 175,000 words in Newton's own hand on the theory and traditions of alchemy, and another 55,000 words of experimental notes.

Something of the familiar Newton appears in all that text. He began his *Index Chemicus* in the early 1680s and put it into its final form in the first years of the 1690s. Ninety-three pages long, with almost nine hundred entries—from Abaranaos Arnald to Zengiufer—it was the most comprehensive listing of alchemical ideas, writers, and concepts ever composed. It reached back to the ancient—perhaps purely legendary—founders of alchemy, traced medieval developments, and duly noted the work of some of Newton's own contemporaries, including Robert Boyle. It was the same kind of list Newton wrote time and again as he organized his thoughts for any research program he undertook.

But then there is *Praxis*—"Practice"—completed at the same time as the *Index* in its final form. *Praxis* records what Newton thought his experiments meant. There, straightforward laboratory records ("antimony melted wth tin 5pts in calido [heats, melts] wth mercury not very difficulty; wth 8 pts it amagalms very easily") evoke passages like this: "The rod of Mercury reconsciles the two serpants & makes them stick to it ... wch bond is Venus."

Or this: "This salt or red earth is therefore Flamels male Dragon wth out wings, for after it is thus extracted out of its native earth it is one of the thre substances wch the Sun & Moons bath is made."

Or this: "Tis the minera of Gold even as o^r Magnet is ye mineral of this o^r Challeybs. . . . T'is a spirit highly volatile, or ffiery Draon, our infernal secret fire."

Nonsense, seemingly, the stuff of fever dreams. That is what (more or less) the syndics of Cambridge University Library concluded in 1888 when they declined to accept the Earl of Portsmouth's donation of a trove of Newton's alchemical writings as "of being very little interest in themselves." And yet the committee did accept much of what Portsmouth offered, including the *Index Chemicus*. The distinction was simple: the real Newton, the official Newton, compounded antimony and mercury in precise proportions and carefully wrote down his results. The other Newton was an embarrassing uncle to be kept in the attic lest he walk unsteadily down Trumpington Street, muttering just a little too loudly about wingless dragons and infernal fire.

Except, of course, the chemical Newton of the lab notebooks and the alchemical Newton who pondered the bathing habits of the sun and moon are the same man. That single thinker was not, almost all of the time, a lunatic. The language in *Praxis* appears opaque, even bizarre, only when it is extracted from its context. Newton in 1693 remained deeply committed to the alchemist's duty to prevent the vulgar from gaining access to knowledge too powerful to be trusted to just anyone. *Praxis,* handwritten, never destined for publication, demanded that its few, carefully selected readers share that deep immersion in the practice and private language of the alchemical tradition. For those who could read past the menstruum of Venus and the salts of wise men, the document does reveal fundamental insights into what Newton believed he knew in the late spring of 1693.

That June, Newton was convinced he had discovered something of transcendent importance. It had been sparked by the hint Boyle had left for Locke to find. Boyle's work implied that there was a process, what alchemists called a kind of fermentation, that could take a base mixture charged with a seed of gold and transform the whole mass into precious metal.

Once Newton jettisoned his lingering skepticism, he began to chase down Boyle's hints about this fermentation, termed the "wet way" of alchemical action. As he felt himself getting closer, his already furious pace intensified, until he was seemingly chained to his furnace, working and reworking his experiments in a reprise of the alchemical frenzy Humphrey Newton had witnessed almost a decade before.

Then he rested. In one of the last passages of *Praxis,* he reported what he had done. He listed a dizzying series of steps, interactions involving sulfur, mercury, and several other compounds. The procedure produced a range of products—a black powder that Newton described as "our Pluto, ye God of wealth o^r Saturn who beholds himself in the looking glass"; then, later in the sequence, the "Chaos . . . that is ye hollow oak"; and later still, "the blood of the Green Lyon." Each step, each chemical product carefully concealed behind this fantastic imagery, took Newton closer to his actual goal, the ability to make the final product that could drive the transformation of matter from one form into another. At last, he wrote, he succeeded, forging the legendary "stone of the ancients."

With that, Newton allowed his most extravagant language to fall away and simply recorded what happened next. "You may ferment them wth ☉ [gold] & ☽ [silver] by keeping the stone & metal in fusion together for a day, & then project upon metals." This stage, "the multiplication of the stone in vertue," created a kind of catalyst, the ultimate goal of millennia of alchemical investigation. Then, as Newton wrote, a touch of color returning to his prose, "You may multiply it in quantity by the mercuries of wch you made at first, amalgaming ye stone wt ye ☿ [mercury] of 3 or more eagles and adding their weight of ye water, & if you desgine it for mettalls you may melt every time 3 parts of ☉ wth one of ye stone . . . *Thus you may multiply to infinity*" (italics added).

The philosophers' stone. Power without limit—and knowledge too. It was the alchemist's dream, realized at last. *Praxis* ends with a discussion of whether his newly formed philosophers' stone was the "quintessential matter or Chaos out of which man and all yᵉ world was made." Newton now possessed, or thought he did, what he had looked for with such hunger throughout his decades of alchemical hard labor: a direct connection to an omnipotent God. He had pursued the same goal in all his work, of course. To paraphrase Albert Einstein, Newton wanted to know what choices God made when He created the world. More deeply, he wanted to understand what comes next—what the divinity is doing now in the physical cosmos of space and time. And here, at last, he thought he had the answer: in multiplying gold at the laboratory bench he had achieved the ultimate act of *imitatio dei,* an imitation of God's will in the world. It was the moment when Newton, and Newton alone, peered into the mechanism of divine action this side of heaven.

This vision he kept to himself. One or two other men may have read a version of *Praxis,* though it is not known whether he showed the final version to anyone. With Boyle dead and Locke sworn to silence, Newton's result—whatever he had truly achieved—remained his and solely his to ponder.

9

"Sleeping Too Often by My Fire"

FOR A MONTH or so, Newton seems to have persisted in the belief that in virtuous imitation of God he could transmute base matter into something much more precious. But he never finished *Praxis*. After he wrote of his apparent success in making gold, he added a comment on two earlier alchemists and then simply stopped, practically in midthought. He never returned to alchemy with anything like the same intensity. The regular spring and autumn tours of duty in the laboratory ended after the middle of 1693. He performed a few experiments in 1695 and 1696, and later wrote some scattered notes on alchemical matters. But nothing after the climax of 1693 comes close to the passion for the work he displayed earlier. Somehow—he never revealed just what convinced him—he understood that the divine secret of the transmutation of matter had eluded him once again. And then, perhaps in response, or at least in sequence, his mind broke.

Newton's delirium overwhelmed him in the early days of June. On May 30, he began a letter to Otto Mencke, editor of *Acta Eruditorum*, Europe's leading scientific journal. The draft ends in midsentence, the bare frame of a question: "*Quid . . .*" Then silence, for almost four months.

In that time, the near mania of the last days of his alchemical passion gave way to blank, black absence. Newton left almost no hints of what actually occurred during the missing months, but he let slip an admission of sleeplessness, of "distemper." He said that he was unable to remember what he had thought or written days or weeks before. The impression—and that's all it is—is of classic depression, misery deep enough to consume one's sense of self. Newton reemerged in September, but his first attempts to reconnect with a wider world merely added to the intimation of something terribly awry in what was supposed to be the greatest mind of the age. On September 13, Newton sent a sad, bitter note to Samuel Pepys, who in his term as president of the Royal Society had authorized the publication of the *Principia.* "I never designed to get anything by your interest," Newton wrote, "but am now sensible that I must withdraw from your acquaintance and see neither you nor the rest of my friends any more." Pepys's sin, apparently, had been to commend him to the reviled King James—a grudge nursed fully six years after the Stuart King had fled the country—but Newton let slip the deeper issue. "I . . . have neither ate nor slept well this twelve month," he wrote, a backhanded admission of the ratcheting stress of his alchemical crusade. Worse, he wrote, "I am extremely troubled at the embroilment I am in"—which had cost him "my former consistency of mind."

The letter to Pepys suggests that Newton was lucid enough to recognize his own distress and to assign reasons for it. But three days later, he sent another letter, to John Locke, that bespeaks true paranoia: "Being of the opinion that you endeavoured to embroil me wth woemen," Newton wrote, "& by other means, I was so much affected with it that when one told me you were sickly & would not live I answered twere better if you were dead."

A month later, however, Newton wrote again to Locke, as if in acknowledgment of—and apology for—a temporary derangement. "The last winter by sleeping too often by my fire," he wrote, "I got an ill habit of sleeping & a distemper wch . . . put me further out of order." Thus, "when I wrote to you I had not slept an hour a night for a fortnight together & for 5 nights together not a wink." He seemed to be asking forgiveness, though for what he was not sure: "I remember I wrote to you," Newton confessed, but what he had said, "I remember not."

In the meantime, scholars on the Continent had begun to pick up hints that something was wrong, and the gossip mills inflated the disaster. Rumor reached Christiaan Huygens that Newton had lost more than a year to what was called frenzy, a condition attributed both to overwork and to a fire that was supposed to have destroyed Newton's laboratory and a portion of his papers. Huygens told Leibniz, and Leibniz talked to his friends. With each telling, Newton deteriorated, until by 1695, Johann Sturm, a professor at the University of Altdorf, reported back to an English correspondent the story of the fire, now so enlarged as to consume not just a shed and some notes, but Newton's house, library, and all his worldly possessions. Sturm described what seems to have been the Continental consensus: the greatest natural philosopher of the age had become "so disturbed in mind thereupon, as to be reduced to very ill circumstances."

The fire seems to have been a pure invention, a nicely mechanical cause to explain how the author of the *Principia* could descend into a delirium so consuming that (again, in rumor) he failed to recognize his own book. But even if his intellectual rivals leapt a little too eagerly to the conclusion that the giant in their midst had fallen, the fact of Newton's collapse remains.

What had actually happened? Newton's derangement was certainly connected, by time, place, and emotional logic, to the fate of his alchemical investigation. But there was another element in the mix, one that Newton never discussed: the end of his nearest approximation to romantic love.

A wealth of myth has sprung up around Newton's emotional life, and the possible existence of a sexual one. He was an irascible man, a great hater. Hooke and Leibniz were only the most famous of those he detested. He did not make friends easily, especially as a young man, and seemed not to mind the resulting solitude. He was not much of a joiner. He never married. Without doubt, he was a prude. Hence the legend of the bloodless semidivine hero of the mind—as Halley put it, of the figure whom "nearer the gods no mortal may approach."

But for all such adulation then and since, the real Newton was a human being, capable of great feeling—affection as well as enmity, loyalty as well as implacable disdain. At least once in his life he demonstrated that he could go so far as to feel a bond with another person stronger than mere friendship. That man was a young Swiss mathematician, Nicholas Fatio de Duillier.

Fatio, then just twenty-five years old, probably met Newton at a Royal Society meeting on June 12, 1689, at which Huygens was scheduled to speak. Newton had come to hear the man who was his nearest intellectual equal, and at some point in the course of the evening he was introduced to the younger man.

They got on from the start. Less than a month after the Royal Society encounter, Fatio joined Huygens to accompany Newton to Hampton Court to seek King William's patronage for the Englishman who had written the *Principia*. The expedition came to nothing, but the fact that Fatio was included in the company was a measure of how much pleasure his company gave Newton.

There was much to like. An undated portrait shows a handsome, even striking young man with large, luminous eyes and a flirtatious half smile. He was more than a mere beauty, however. He was enough of a mathematician to catch at least one error in the *Principia*, and apparently considered editing a second edition of the book himself. He was just wise enough not to overplay his hand. As he admitted to Huygens, for all of his pride in catching a mistake or two in the great work, "I was frozen stiff when I saw what Mr. Newton has accomplished."

Even better, he was a born disciple. Early in their acquaintance,

Newton, having only just returned to Cambridge, wrote to Fatio of his plans to go back up to London to see him. Fatio replied, "I had a design to go see you at Cambridge; but you send me word you will come hither, which I am very glad of." More than glad, it turned out, for Fatio was trying in his own way to emulate his hero, going so far as to attempt a new interpretation of Newton's conception of the mutual attraction between bodies: "My Theory of Gravity," Fatio wrote, "is now I think clear of objections and I can doubt little but that is the true one." But who is he to say so, when the master is at hand? "You may better judge when you see it," Fatio told Newton, with all due deference. (In fact, another of Newton's friends, the mathematician David Gregory, reported, "Mr Newton and Mr Hally laugh at Mr Fatios manner of explaining gravity.") He declared himself not only Newton's "most humble and most obedient Servant," but one who was "with all my heart" devoted to Newton's service.

Newton was smitten. After the summer of 1689, he made sure that on his next visit to London, he took rooms in the same house with Fatio. The two stayed together for as much as a month in March 1690, Fatio acting as Newton's secretary, transcribing revisions to the *Principia*. Already, less than a year into their acquaintance, Newton devoted more time to the company of a single person than he had ever done before or would do again.

That first, intense involvement ended fairly quickly. Fatio returned to the Netherlands, and Huygens in June. He stayed away for fifteen months. With distance between them, he seems to have withdrawn from the pull of Newton's attention—at least, Newton complained in a note to Locke that Fatio had remained silent for months.

Nonetheless, when Fatio returned to England in September 1691, Newton immediately traveled up to London. Over the next year and a half, the two men remained in regular contact, when Newton came to the capital and at least once when he received Fatio in his rooms in Cambridge. They still talked natural philosophy, but the relationship and its overt focus shifted. Newton began to honor Fatio with secrets he revealed only to Boyle and Locke and perhaps a few others. In the clearest expression of the value he placed on this one young man, he opened his most intimate thoughts, enlisting Fatio as his alchemical acolyte.

Throughout the early 1690s, Fatio eagerly followed his mentor into the secret corners of the Newtonian cosmos. Newton composed his last version of *Index Chemicus* and then *Praxis* during the years of his most intense affection for Fatio, and both were probably at least partly intended to be personalized texts, composed for a beloved disciple by the first and greatest truly quantitative alchemist in history.

Few details survive of Fatio's own alchemical research, but in the spring of 1693, he sent Newton an account of an experiment with mercury for use in a reaction with gold. The report reflects Newton's insistence on sound laboratory procedure. Fatio specified the instruments used: "a wooden mortar that hath a pestill allmost so big as to fill exactly the mortar"; he notes the efforts to reduce impurities in the preparation; he describes as exactly as possible the sequence of events as the mixture changes color and sprouts "a heap of trees out of the matter."

This was flattery by imitation. Fatio did the same across the range of Newtonian passions — besides his dabbling in alchemy and his rather feckless attempt to make sense of gravity, he followed Newton into scriptural exegesis, writing about the interpretation of prophecy, the temptation of Adam and "the serpent being there the Roman Empire." Such earnest prose on Newton's subjects, all in echoes of Newton's own rhythm and voice! It is a familiar spectacle, an ancient one: a comely, talented youth craving the attention and submitting to the loving instruction of an older, powerful man.

It did not last.

Fatio wavered first. He fell ill in the autumn of 1692, and painted a dire picture for Newton: "I have Sir allmost no hopes of seeing you again," he wrote. He had picked up a terrible cold on his last visit to Cambridge, and it had settled in his lungs. Disaster followed disaster—something that felt like the rupture of an ulcer in his lungs, a fever, deepening mental confusion. This could really be the end, Fatio warned, though he admitted he had not yet seen a doctor, who "may save my life." Should the worst happen, a friend "will acquaint you with what may befall me."

Newton replied in a frenzy of advice and sympathy. On September 21, he wrote, "I . . . received your letter wth wch how much I was affected I cannot express." He gave orders: "Pray procure ye advice & assistance of Physitians before it be too late." No excuses were permitted: "if you want any money I will supply you." All this, "with my prayers for your recovery" from "your most affectionate and faithfull friend to serve you."

Fatio wrote again the next day, acknowledging his friend's concern: "I give You Sir my most humble thanks again, both for your prayers," he wrote, "and for your kindness to me." Apparently his earlier report had been a false alarm, an eruption of Fatio's gift for self-dramatization. But there seemed to be another reason for his shift in tone, now not only markedly less urgent than Newton's but with none of the ardor of the eager early days of their friendship. Fatio was polite, even graceful, and signed off as his friend's "most humble, most obedient and most obliged Servant." But that was simply the conventional formula, more polite than felt.

Newton caught the shift. For the only time on record, he acknowledged dependence, fear, a kind of desire. He began to beg. He suggested that London's evil air might be damaging his friend's health. The solution: come to Cambridge, "so to promote your recovery & save charges till you can recover I [am] very desirious you should return hither." Newton offered money, a home, care, whatever it took to secure the health—and favor—of his companion. Fatio resisted all temptation, but left Newton with a hint of hope. He wrote back that he had decided to return to Switzerland, inquiring coolly, "I should be glad Sr to know whether your occasions will possibly bring you to London before I go away."

But then he intimated that he might yet return to England, and if he did, he could settle in Cambridge. "If You wish I should go there," Fatio wrote, "I am ready to do so"—having, he added, other reasons for such a move than merely saving money. Newton took heart from that show of willingness, and accepted the news of Fatio's impending travels relatively calmly. "I must be content to want your good company, at least for some time," he said, taking his leave "wth all fidelity" until they could meet again.

There the exchange rested. The two traded a few letters about minor matters, a box of rulers Fatio had left behind, some books, and so on. Newton continued to press for a reunion in Cambridge, and Fatio continued to alternate suggestions that he might come to his friend with an always formidable propensity for self-absorption. (He wrote, for example, of his plan to undergo an operation for hemorrhoids, "to be free of an excrescence ... very troublesome to me.")

Then, abruptly, Fatio announced a change of plans. In May, five months after he first hinted that he might settle with or near Newton, he announced that he had made "new acquaintance ... a good and upright man." This new friend was another alchemist, an expert in the creation of mercury compounds. Fatio wrote that this adept could cure consumption with a potion "which he giveth for nothing." So, he assured Newton, there was no need to worry about his health anymore, for that good man would care for him without charge. "I have not now any need of money Sir, nor of your powders, but I thank you kindly for both."

Just in case Newton missed the point, Fatio wrote again of his new friend's marvelous remedy, announced his own plan to study medicine, so as to be able to sell the potion himself—and then had the boldness to suggest that Newton might want to partner with him in the enterprise. There was no mention of Cambridge, nor of rooms to be taken side by side, nor of the alchemical work that master and disciple could undertake. Fatio did allow that were Newton to come to London, he would be happy to see him. Failing such a meeting, he retained "all manner of respect" for the man to whom he had formerly pledged the whole of his heart.

To this, Newton had no reply. The next known document in his own hand is the one he abandoned on the question "*Quid . . .*" "What?"

Was Newton's heart broken? Maybe. Certainly, there is no parallel in the rest of his long life for the urgency he allowed himself to reveal in his letters to Fatio. His panic at the hint he might lose the object of that affection had no precedent either. Newton was more accustomed to dismissing men from his acquaintance than pursuing them.

Were Fatio and Newton lovers? No one knows. Probably not, if the rest of Newton's life is any guide. Except for his exchanges with Fatio himself—always written in the perfectly respectable terms with which men expressed affection then—there is nothing that can be seen as a love letter anywhere in his correspondence. The only more or less explicit traces of erotic longing Newton admitted throughout his life appear in his catalogue of sins as a young man. Along with lying about a louse to allowing his mind to wander during chapel to harboring murderous thoughts about his mother, he admits to "having uncleane thoughts words and actions and dreamses" and turning to "unlawful means to bring us out of distresses."

And that's it, just about all that connects Isaac Newton to sexual desire. His accounts contain no debits for bawdy houses, although they list several for taverns and drink. His letters mention nothing of physical need. His private papers add nothing to the image of a mostly sexless creature. Whether or not the living man who woke up alone each morning and laid his solitary self down at night felt any hunger for another's touch, man or woman, he seems never to have admitted as much—to himself or to anyone else.

In any event, the question of Newton's notional sex life misses the point. His letters to Fatio, and the younger, more beautiful man's replies, portray a clear and rather sad longing for some kind of intimacy—an emotional closeness, whether or not bodies were involved. For the short span of a year or two, Newton seems to have felt such a connection.

Fatio disappeared from view soon after he dismissed Newton. He eventually turned up at Woburn Abbey, tutoring the Duke of Bedford's children. He never regained his status as a rising star in European intellectual life. He never produced any original mathematics. Ultimately he became something of a pathetic figure, prone to religious manias and sponging cash from former friends—including Newton, whom he touched for thirty pounds in 1710.

But in the immediate aftermath of Fatio's flight, Newton suffered more. No one can now discover precisely why Newton broke down. He had some history of melancholy, from his childhood plaint ("I know not what to doe") to the prolonged periods in the 1670s during which he withdrew almost completely from public scrutiny. But the sequence of events is there. Newton's crisis followed directly on the collapse of his feelings for his friend. Put that together with his recognition of alchemical failure, and May and June of 1693 become scorched earth, the ruin of all hope. Fury, sorrow, silence—each of these are plausible reactions to such devastation, and Newton endured them all.

The summer passed. Newton gave no sign of noticing. September came, and he began to speak bitter, unhappy words—those blows flung almost at random against Pepys and Locke. Pepys did not respond, consciously choosing not to notice the disintegration of his great friend. Locke answered, hurt, but still declaring that "I truly love & esteem you & . . . I have the same good will for you as if noe thing of this had happened."

October turned, and Newton proceeded with the slow reorganization of his disordered mind. He apologized to Locke, and his friend forgave him. In November, Newton completed the letter he had abandoned in June. Finally, canny old Pepys reached out, with never a mention of any odd behavior on the part of his friend. Instead, he asked a technical question of great interest to gamblers: who had the best odds in a particular game of dice.

Newton understood the intent behind the message. He wrote back, "I was very glad . . . to have any opportunity given me of shewing how ready I should be to serve you or your friends upon any occasion." He added that he wished he could perform some more important task, but nonetheless analyzed the issue at hand, explaining how Pepys should place his bets.

From there Newton's recovery strengthened. He resumed the company of his friends and worked closely with several younger colleagues, David Gregory and Edmond Halley especially. Fatio became a man he used to know. They did correspond infrequently—an exchange in 1707 centered on Fatio's enthusiasm for an apocalyptic religious revival being preached in London. But this was action at a distance. None of Newton's later letters to Fatio suggest a shred of pleasure in the other's company.

The winter of 1693 came, then passed. Through the spring and summer of 1694, Newton concerned himself with this and that. He wrote a long memorandum on the proper education of boys. He dealt with nettlesome tenants on the land he had inherited from his mother. He made some notes on problems in calculus. He began what would be a years-long effort to come up with a complete explanation for the motion of the moon—the notorious "three-body problem" involving the interaction of the earth, sun, and moon.

The lunar calculations were good work, though Newton ultimately concluded (correctly) that he had failed to find a solution. He remained a brilliant mathematician. His skill was put to the test in 1697, when Johann Bernoulli published a pair of problems—a challenge that aimed at the most prominent mathematicians of the day. Newton received a copy of the second of the two on January 29, at four in the afternoon. By four the next morning, he had solved both. He sent Bernoulli his calculations unsigned. Bernoulli was undeceived, recognizing the mind behind the work *"tanaquam ex ungue leonem"*—"as the lion is recognized by his print."

But good or great, such efforts were trivial compared with what Newton had achieved before. It is unfair to ask for two *Principia*s from any man. He did still produce prodigious amounts of work, but increasingly his writing centered on history, biblical criticism, the analysis of ancient prophecy. His breakdown probably helped impel the shift in emphasis, but the simple fact is that time was passing. Very few creative scientists perform at the highest level for decades on end, and Newton had been at the leading edge of discovery since his early twenties. On Christmas Day, 1694, he turned fifty-one.

The year turned again. The academic calendar rolled on. Newton remained in residence at Trinity College, if not actually bored, then underemployed. There were whispers something might be coming his way, some plum post that would at last pry him loose from a university whose life absorbed him less and less. Nothing materialized, but in September 1695, an odd message arrived from London. It was a request on a matter completely outside his usual competence:

Would Isaac Newton kindly provide his thoughts on a matter of national importance? What should the nation do about the worsening shortage of silver coins?

Part IV

The New Warden

IO

"The Undoing of the Whole Nation"

WILLIAM LOWNDES, SECRETARY of the Treasury, had a problem that had been growing worse for years. For at least half a decade, it had been evident to anyone paying attention that there was something wrong with England's money. Specifically, there wasn't enough of it. The silver coinage, all the denominations from half-groats (two pence) to crowns (five shillings), was evaporating. From the late 1680s to the mid-1690s, the supply of these coins—the basic units of exchange for the daily business of the country—shrank year by year. By 1695, it was almost impossible to find legal silver in circulation. Something needed to be done, and it was Lowndes's job to suggest the proper course of action.

He sought help. In September 1695, he wrote a letter asking advice of England's wise men. Some were obvious choices. John Locke had written a series of papers on money and trade in 1691. The architect and polymath Sir Christopher Wren had extensive experience with both government and budgets in his role supervising the rebuilding of London's churches and St. Paul's Cathedral after the Great Fire of 1666. Charles Davenant was one of England's leading writers in the field just beginning to be called political economy, and he had served as an excise official, administering England's tariffs. Most of the rest of the men Lowndes contacted were more such eminences: the banker Sir Josiah Child, a major shareholder in the East India Company; a lawyer, John Asgill; and Gilbert Heathcote, a governor of the newly formed Bank of England. But Newton?

The *Principia* had established Newton as the smartest man in England and thus a natural figure to be called upon at a time of national crisis. That he had no knowledge of government finance or experience of the market was hardly a handicap. The beginnings of the modern economy predate the emergence of economics as a formal discipline and of that special class, economic experts. And so it happened, without any apparent hesitation, that England's greatest natural philosopher first turned his mind to the problem of money.

The Mint and the Treasury had been wrestling with the damage done to the currency by coiners and clippers since the early 1660s. But at about the time the Stuarts fell and William rose to the throne, a new threat emerged: the trade in silver from England to Amsterdam, Paris, and beyond. The exchange was driven by a difference in the price of silver against gold in London, compared with the prices on the Continent. Simply: you could buy more gold in France with a given lump of silver than the same weight of English minted coins could purchase in London. There was no shortage of clever operators who figured out the arbitrage opportunity: collect silver coins in England, melt them down into ingots, ship them across the Channel, buy gold, and then use that gold to buy yet more silver back home. It was the nearest thing imaginable to a financial perpetual motion machine.

By 1690, within two years of the coronation of William and Mary, the outflow of silver coins became acute enough to provoke a parliamentary investigation. Several members of the Worshipful Company of Goldsmiths—the guild governing dealers in precious metals—petitioned for aid to prevent what they said was the ruin of their business. In the preceding six months alone, they claimed 282,120 ounces of silver had been shipped out of London to metal dealers in France and Holland—enough to strike at least 55,000 pounds sterling of minted money, more than ten percent of the total silver coinage struck at the Royal Mint in the previous five years. Who was responsible? Never one of their own company, to be sure! Instead, the goldsmiths implicated foreign metal dealers, especially those ubiquitous and useful villains the Jews, "who do any thing for their profit."

A committee led by Sir Richard Reynell was formed to pursue the goldsmiths' charges, and on May 7, Reynell rose in the House of Commons to report the results of the inquiry. The petitioners' claim was a fact: silver was indeed abandoning the kingdom. There was no mystery about the reason. The difference in value of English silver bullion on the continent of Europe, compared with the face value of the full-weight legal shillings coined out of each ounce, was not much—about one and a half pennies per ounce of silver. But that was profit enough, according to the parliamentary investigators, to make it worthwhile for traders to turn English money into ingots to be sold across the Channel.

Reynell was a little more temperate than the petitioners in his assignment of blame. While "the Jews, for their Profit, exported [silver] in very great quantities ... to the utter Ruin of the working Goldsmiths," Reynell admitted that there lived "English, as well as Jew, who for their Advantages, would doubtless melt down our Crown Pieces, &c and sell for Foreign Silver to the Undoing of the whole Nation for want of Money, unless a present Remedy were found to prevent Exportation of any Silver or Gold."

Making matters worse was the other half of England's currency debacle: the existence of those two parallel coinages— the old, hand-struck, pre-1662 money and the newer, heavier, machine-made pieces. Bad money was driving out good. The machine-made money, precisely weighed and secure, would never circulate so long as debased coins passed for the same face value. The great Victorian historian Lord Macaulay later reported that as the crisis reached its climax, the Exchequer took in no more than ten good shillings in a hundred pounds of revenue—one out of every two thousand coins. Macaulay wrote, "Great masses were melted down; great masses exported; great masses hoarded; but scarcely one new piece was to be found in the till of a shop or in the leathern bag which the farmer carried home after the cattle fair." The resulting crisis was, he wrote, much more serious than the misgovernment of Charles and James. "It may well be doubted that all the misery which had been inflicted on the English nation in a quarter of a century by bad Kings, bad Ministers, bad Parliaments and bad Judges was equal to the misery caused in a single year by bad crowns and bad shillings." It did not matter to most Englishmen who ruled in London "whether Whigs or Tories, Protestants or Jesuits were uppermost, the grazier drove his beasts to market; the grocer weighed out his currants; the draper measured out his broadcloth; the hum of buyers and sellers was as loud as ever in the towns." But when "the great instrument of exchange became thoroughly deranged, all trade, all industry were smitten as with a palsy. The evil was felt daily and hourly in almost every place and by almost every class."

Gold guineas could still be found, costing about thirty shillings at a goldsmith's bank. But a pound of beef at Spitalfields market went for about three pennies in the spring of 1696. A gallon of beer ran a shilling or less. A laborer's daily wage was thirteen pence or so. As the small silver coinage that was the engine of daily life disappeared, trade suffered, and then almost stopped. "Nothing could be purchased without a dispute," Macaulay wrote, and "the simple and the careless were pillaged without mercy." The Mint had produced nearly half a million pounds' worth of silver currency between 1686 and 1690. But so much silver poured out of England in the next five years that the Mint could find almost none to coin, producing just over seventeen thousand pounds between 1691 and 1695.

Reynell and his colleagues confirmed the facts of the crisis, but "though the Committee found the complaint of the Petitions very just and the Inconveniences to the Kingdom very great, they could not agree of a way for preventing the same." A law on the books prohibited the melting of minted coin, but as long as English silver was worth more as bars of metal than the Mint said it was as crowns or shillings, England's hard cash would continue to vanish down the Thames.

Nothing was done in that session of Parliament, or in the next, or the next. All the while, as Macaulay put it, "the coins went on dwindling and the cry of distress from every county in the realm become louder and more piercing." For five years, arguments about the crisis raged across London. Finally, the one man with the power to demand action found himself personally in danger for lack of good silver coin. In July 1695, King William commanded a mixed army of English and Dutch soldiers besieging the French in the fortress city of Namur, in present-day Belgium. The campaign was part of William's grand strategic attempt to check Louis XIV's power in Europe and beyond. The two sides had already been fighting for seven years and would continue for more than a century, in what Winston Churchill would correctly term a world war. But at this particular moment, William faced the prospect of being defeated not by force of arms but by lack of cash to keep his army in the field.

The difficulty stemmed from a change in the way Europeans made war on each other. In a land campaign, the contending armies engaged in a series of assaults on fortified positions. It was slow, indecisive, positional warfare, dominated by combat engineers and the artillery, interspersed with sudden bursts of intensely bloody face-to-face combat whenever the cannons managed to blow a breach in the opponent's defenses. Both sides responded to the resulting stalemate by ramping up their armed forces. Louis's France, at war for decades, had already boosted the size of its standing army. The English followed suit. From just 25,000 men under arms at the beginning of the war, William's army grew to about 100,000 by the mid-1690s.

Supporting militaries on that scale forced radical change, not just in the types of battles such oversized forces could fight, but in the way governments and nations organized themselves to pay for their ambitions. In England, the necessary changes turned on the conditions attached to William's rise to the throne. He held power not by hereditary succession but by the gift of an elected lawmaking body, the Convention Parliament. It was a carefully limited grant: the elected members held on to the power of the purse. William himself received a salary from the state, thus becoming the first monarch to serve as the most prominent member of what was just becoming a professional civil service.

And most of what that nascent civil service did was to figure out ways to extract from the English people the money needed to run the ever more ambitious national government. William's revenue-collecting bureaucracy tried land taxes, customs duties, excise charges. In 1691, Parliament passed a bill authorizing a levy of more than 1.6 million pounds to pay for "the Carrying on a Vigorous Warre against France." As a sign of the growing reach of the government, it appointed tax commissioners in cities and counties all over England and Wales—among them, for "the University and Towne of Cambridge," a Mr. Isaac Newton. The government borrowed as much as it could, much more than any previous English administration. William's ministers created a whole new kind of debt, an early form of government bonds, in 1693, raising a million pounds in one issue, more in another. It still wasn't enough to feed and arm the troops needed in the field, so in 1694, Parliament chartered the Bank of England. By the end of 1695, the Bank had already lent the government 1.2 million pounds.

Even such enormous sums were not enough. By the mid-1690s, spending on the war exceeded government tax income. Worse, the export of good silver money as bullion, and the assault of the clippers and coiners, meant that the government collected much of its income in coins so debased that no private trader—and, crucially, no foreign banker—would accept them at face value. By 1695, the exchange rate for English silver currency in Amsterdam was dropping steadily. By midsummer, the cost of the war had affected both high finance—the ability of the government to raise large sums through loans—and the basic supply of hard cash, all that English silver that had vanished from the coinage. Put the two together, and William's army was perilously short of money.

The crisis could not have come at a worse moment. Taking Namur would be both a strategic and a symbolic victory, but not if William could not press his campaign there. Absent a ready supply of money from London, it fell to the army's paymaster, Richard Hill, to find some cash fast. He traveled to Brussels to solicit funds from the rich banking community there, but it took him months to secure a loan of 300,000 florins, explicitly because of the state of English government finances. The money did reach the army before it dissolved into rabble, and Namur fell on the fifth of September, but the war dragged on. For reasons of state, perhaps, and certainly of personal vanity, Louis XIV would not enter serious peace negotiations in the wake of a very public defeat. So, as the campaigning season ended in late 1695, it was clear that the war would resume the following spring—unless one of the combatants happened to go bankrupt in the meantime.

The implications were obvious, certainly to William and his government. If England were to continue to fight, it needed a stable currency. When William opened the session of the House of Commons on November 26, 1695, he almost begged its members to respond to the currency crisis.

He began with apparent diffidence, acknowledging that it was a "great Misfortune, That, from the Beginning of my Reign, I have been forced to ask so many, and such large, Aids of my People." But, he warned, there was to be no relief. "I am confident you will agree with me in Opinion," William said, "That there will be, at least, as great Supplies requisite for carrying on the War by Sea and Land this Year, as was granted in the last Session"—more, in fact, for "The Funds which have been given, have proved very deficient." William acknowledged the "great Difficulty we lie under at this time, by reason of the ill State of the Coin." Fixing that problem would cost yet more money the government did not really have, but this was "a matter of so general concern, and of so very great importance, that I have thought fit to leave it entirely to the consideration of the Parliament."

It was a neat display of rhetorical jiu-jitsu. The King modestly deferred to the Commons—he was, after all, monarch by grace of Parliament's vote—to figure out who should suffer to fund his unpopular war. But the question remained: what could the government actually do to prevent the sale of English silver to the highest bidder?

Hence Lowndes's plea for help, and the answers that came in from the good and the great—among them Isaac Newton.

II

⸻⸺∞⸻⸺

"Our Beloved Isaac Newton"

TO NEWTON, WORKING through the problem Lowndes had set, one fact was obvious. Though he did not put it in quite this language, it was clear that currency criminals were rational actors responding to an uncomplicated set of incentives. Silver clippings represented pure profit, as was the margin harvested abroad for full-weight minted shillings. Human beings would continue to take those gains unless compelled by coercion or a change in the marketplace. The problem was as straightforward as the simplest of equations.

Newton also understood that force alone could not eliminate the smuggling of bullion, given that the crime of clipping persisted despite the death sentence it carried. So he turned his attention to the source of profit in the illegal silver trade, and came up with two measures that could destroy the elementary economic logic behind the assault on England's coins. First the nation had to get rid of its old, worn, increasingly debased currency. To do so, Newton and many others recommended a complete recoinage. All of England's silver money, old and new, was to be called back in to the Mint, melted down, and remade into a single, consistent, edged issue. That step alone would mostly solve the clipping problem. With no more hand-hammered, smooth-edged money in circulation, it would become exceedingly difficult to shave much metal from the new coins.

But without a shift in the ratio of weight to face value of the new coins, reminting England's money would not curb the relentless flow of silver across the Channel. To solve that problem,

Newton argued, it was essential "to make Milled money constantly of the same Intrinsick & Extrinsic value, as it ought to be and thereby to prevent the Melting or Exporting it." That is, instead of two different sources of the value implied by a coin—the "intrinsic" market price of its metal and the "extrinsic" value imparted by the stamp of the head of a monarch that transformed an ordinary disc of metal into legal tender—these separate measures must be brought into agreement. With silver and gold both being used as money, this meant altering the relative value of the two. In this case, when England's silver bought more gold on the Continent than it could buying guineas at face value, that meant lowering the amount of silver per shilling—making Dutch or Spanish gold more expensive as counted in English silver money. Such a devaluation, performed correctly, would eliminate the price differences exploited so successfully by the currency buccaneers.

Lowndes, the leading public figure arguing for devaluation, welcomed Newton's reasoning and support. He still found it hard to make his case, because at its core was a radically modern thought: the King's imprimatur was a mere fiction and not the working of a kind of magic that determined the absolute worth of a given piece of silver. By Newton's logic, the word "shilling" could be thought of as no more than a convenient way to express what a given amount of silver bullion was worth as a commodity. In that view, units of currency—shillings, half-crowns, guineas—could not be absolute statements of value, extensions of the divine authority of kings. Instead, they were relative claims of the prices of quantities of metal—of anything—and those values could change with every shift in conditions in the real world.

Thus, lurking within the argument for devaluation lay a genuinely unsettling idea. Money need not be seen as merely a thing, a tangible object jangling in one's purse. It could be understood as a term in an equation, an abstraction, a variable to be analyzed mathematically—as in fact skilled traders had been doing more or less explicitly every time they played the markets in Holland against those in London.

Newton himself did not at first grasp the full implications of his analysis. He still had moments when he thought that the government could, on its own, fix a value for England's silver. He told Lowndes that after devaluation, any dealer who offered a higher price for silver by weight than the face value of the same weight of milled money should be jailed "till the Party offending shall give an Account of himself." But the underlying logic of his discussion of the two sources of value led implacably to the conclusion that devaluation was the only way out of England's currency predicament.

This was a thought too far—if not for Newton, then for most of his colleagues. The unquestioned leader of the anti-devaluation forces was John Locke. To be sure, Locke recognized the need to recoin; the miserable state of the clipped coinage was as obvious to him as it was to anyone in England. But apart from melting down old silver to mint new coins, all else—the old weight and face values for each denomination—should remain constant. To do otherwise, he argued, would violate the very nature of money. After all, changing the number associated with a coin, calling a crown-weight piece of silver seventy-five pence instead of sixty, for instance, would not make that coin buy more silver bullion than it had previously. "I am afraid no body will think Change of Denomination has such a Power."

Locke's argument is correct; it is merely another way of stating the fact of devaluation: a devalued silver shilling contains and buys less silver metal than a higher-silver-content one did yesterday. But that was beside the point. The reason silver escaped to Amsterdam was because each transaction brought more Dutch gold than the same weight of silver stamped into shillings and crowns could buy in England. Nonetheless, Locke denied that units of money—shillings or pounds or pistoles, for that matter—could be subject to a market of their own, varying in price just like any other commodity. Lowndes was his chief target, but Locke did not flinch from contradicting his dear friend as well. In direct rebuke to Newton's thinking, he wrote, "Some are of the opinion that this measure of commerce [the currency], like all other measures is arbitrary, and may at pleasure be varied, by putting more or fewer grains of silver in pieces of a known denomination." Not so, he claimed. "But they will be of another mind when they consider that silver is *a matter of nature different from all other*" (italics added). It was, he said, "the thing bargained for as well as the measure of the bargain." To Locke, silver was unique in the material world: alone in nature it was the fixed center around which all else learned its worth.

Newton was right, but Locke grasped what his friend did not. Devaluation was a weapon aimed at the moneyed, and especially the landowning class—those whose rents would fall by the amount of silver shaved from the legal measure of a shilling piece. Since 1691, Locke had defended a permanently fixed monetary system as a matter of social necessity, a guarantor of the stability of the state. Now he argued that devaluation would "only serve to defraud the King, and a great number of his subjects, and to perplex all." In hard numbers, landlords and the government stood to lose twenty percent under the proposal advocated by Newton and Lowndes.

Locke's view won, of course. When on January 17, 1696, Parliament finally approved the recoinage, it stipulated that the new coins retain the old weights. Four days later, King William gave his royal assent to the act.

There was a pause before recoinage began in earnest. Absent some compelling reason to go to London, Newton simply stayed put, as he had for most of the preceding three decades. But on March 19, he received a letter from Charles Montague, Chancellor of the Exchequer, notifying him that the King intended "to make Mr. Newton Warden of the Mint." Montague had been one of the first men Locke enlisted to help find Newton a job. His rise to the chancellorship in 1694, coupled with the timely resignation of the incumbent Warden, had created the opportunity to provide his old Trinity College colleague with a position in London.

Newton could not respond swiftly enough. Trinity's records show that he left Cambridge for London on March 21 to discuss his prospects. Evidently what he found at the Mint's headquarters in the Tower of London satisfied him. The Chancellor had assured him that the Warden "has not too much bus'nesse to require more attendance than you may spare." By April 13, the paperwork was done. William III, "By Grace of God, King of England, Scotland, France and Ireland," confirmed that the office of Warden of the Mint now belonged to "Our beloved Isaac Newton, Esquire."

One week later, Isaac Newton left Trinity College for the last time. His luggage—including his library of several hundred volumes—would have gone ahead, on one of the carts that made regular hauling runs down the London road. For his own journey, he could have chosen to jounce with strangers on one of the early stagecoaches that had just begun to run from the provinces to the capital. More likely, he would have hired a horse, as became a gentleman. He would probably have broken the journey at the inn at Ware, waiting there, just as Chaucer's pilgrims had three hundred years before, for enough of a company to gather to provide mutual protection along the isolated stretch of road that followed, a notorious haunt of highwaymen.

From there, London was no more than several hours away, and with it a new life, unencumbered—or so it was supposed to be—by any mundane claim on Newton's time or brain. There is no evidence that abandoning his Trinity colleagues cost him any pang. Not a single letter survives between him and anyone he left behind.

12

"Stifling the Evidence Against Him"

NEWTON WAS HARDLY the only man to gain by what was finally understood as a genuine national crisis. William Chaloner quickly recognized the movable feast created by war, debt, and the collapse of the currency. The only problem was where to begin, given all the ways to take advantage of the desperate shortage of cash. The most obvious choice was to respond to the demand for an adequate supply of currency, which is why the middle years of the 1690s were (literally) the golden age of English counterfeiting. The tally Newton would make in 1696 found that more than one in every ten coins in circulation was a fake.

But among all the counterfeiters racing to get rich off the crisis, Chaloner was unique now in recognizing that he could use his knowledge of coining to play both sides of the street. This time his scam was far more sophisticated and ambitious than simply betraying the occasional accomplice. His biographer referred to it as "his Double Deception, serving and cheating the Nation." He had in his sights nothing less than the Royal Mint itself.

Chaloner's first shot at the Mint came in a spray of paper. The collapse of the coinage had evoked a barrage of broadsides, pamphlets, petitions to Parliament, even books. The influential economic thinker Charles Davenant opined on how to pay for William's Continental war, and John Locke weighed in with at least three short pieces on the proper response to the failing money supply. But the Republic of Letters did not belong only to the well-connected. In that era of emerging global trade, the empty tills of London's markets presented a genuinely novel phenomenon that could not be solved by a return to the practices of a simpler age, as many of the pamphleteers pointed out. They presented an array of observations (*The Groans of the Poor*) and solutions (*Proposals for Supplying the Government with Money on easie Terms*). By the height of the crisis, it seemed that everyone in London (and beyond) had some strong view about the nation's finances, and the field was open to an astonishing number of the literate who were willing (and able to pay) to set their thoughts in type. This flood of polemic and diatribe was not merely a reflection of the agitation caused by the crumbling coinage; it also provides another way into the daily experience of what has been called, too narrowly, the scientific revolution taking hold in England.

In the living memory of those who made the discoveries — Newton included — paper as a tool of thought and a means of communication had been constrictingly scarce. The first English paper mill was established in 1557, but it almost certainly made only coarse brown paper used for wrapping, not higher-quality white paper suitable for writing or printing. All of England's writing paper came from Italy or France, and twenty-four sheets cost the equivalent of a day's wage for a laboring man. This is one of the reasons that Shakespeare's plays made it into print only after they were widely acknowledged as a grand achievement. About eighty thousand reams suitable for printing or writing — or roughly seven sheets per person — were imported to England in 1623, the year the First Folio was published (domestic production was still essentially zero). Combined with the cost of the printing itself, the risk of committing anything to the press was so high that no rational businessman would risk a print run unless he was sure of his market.

But by the 1690s, paper imports had shot up, and about a hundred English mills were producing paper domestically. Paper and printing equipment remained expensive, which helps account for the small print runs of even the most important books, like the two hundred fifty or so copies of Newton's *Principia*. But ideas conveyed in the abstract, impersonal medium of the page became available in late-seventeenth-century England at levels unthinkable a century before. Since its 1665 debut, England's first continuously published newspaper, the *London Gazette*, had been joined by a cascade of printed works, passing arguments from mind to mind without the need for face-to-face confrontation. The range of an individual voice now extended far beyond the limit of an orator's shout.

The rise of a technology and culture of (relatively) cheap texts could not determine the course of the revolution in science, or in any body of ideas. But it did have an enormous effect on the speed with which it could achieve its impact. One could offer an account of the value of systematic climate measurements, or of the proper way to calculate the flight of a cannonball—or, at a moment of recognized peril, one could take up the problem of the coinage. Hundreds did, with suggestions that were good, bad, ambitious, fanciful, and even criminal. Among them, William Chaloner.

Chaloner first appeared in print in 1694. In a pamphlet titled *Reasons Humbly Offered Against Passing an Act for Raising Ten Hundred Thousand Pounds*, he made the rather modern-sounding argument that it would be a mistake to raise taxes to make good the shortfall in government revenues caused by deficiencies in the coinage. He had plenty of sparring partners on this issue. One man proposed an inheritance tax of five percent—shocking!—while another proposed higher property taxes for the rich. To no one's surprise, then or now, such ideas went nowhere, and Chaloner was not so foolish as to countenance anything so unthinkable. Rather, the alignment of his views with the interests of men who could be useful in the proper circumstances was almost certainly no coincidence.

Even so, it is hard to avoid the flash of unintended comedy here. William Chaloner writing on tax policy is a bit like John Gotti weighing in on Social Security, or the Kray brothers offering their thoughts on the National Health Service. It is a credit to Chaloner's contemporaries that they seem to have been unmoved by his arguments. None of the wilder schemes to raise revenue were taken seriously, and a Parliament composed of propertied men hardly needed a probably illiterate weaver's son and sometime receiver of stolen goods to tell them how to protect their estates. But the work served Chaloner's purposes, since it was, in essence, his warm-up act.

Chaloner took aim at his real target a few months later. This time he tackled a subject on which he did possess expert knowledge, the benefit of which he kindly revealed under the title *Proposals Humbly Offered, for Passing, an Act to Prevent Clipping and Counterfeiting of Money.* The first part of this booklet takes up an odd but unquestionably innovative idea to rescue the ever-shrinking coin supply. Chaloner proposed a swift recoinage to issue underweight coins—a currency that would measure about two-thirds of the legal standard, similar to what Newton proposed. That devalued coinage would, he argued, render unauthorized clipping unprofitable. Chaloner then broke new ground: after a brief interval to drive the amateurs out of business, he suggested, the entire stock of money should be recalled a second time, melted down, and remade once more at full weight.

It was a clever-sounding notion—but wholly unrealistic, given the cost of even a single recoinage and the Mint's incapacity for the kind of efficiency Chaloner's proposal demanded. No matter. Chaloner was not actually trying to solve the currency problem; he was advertising himself as an expert on the coinage, a man to notice, and perhaps to use—a point the second part of the pamphlet drove home.

Here Chaloner welcomed his readers into the daily life of a counterfeiter. "All Coyning is done either by Casting or Stamping it," he informed his audience. Skilled practitioners often used silver at or close to Mint standards of fineness, making their profit by striking slightly undersized copies of the real stuff. They needed specific tools: those who cast counterfeits used sand molds held in heatproof flasks, while "Stamping mony is Principally done by the use of Flatting Mills and Sheers." Using casting techniques, Chaloner claimed, "in a Daies time one Man can make 100 [pounds]," while, in somewhat more labor-intensive fashion, "By the use of the Flatting Mills, the Coyners of mony do Flat Silver, which they afterwards Stamp, and with the Sheers cut it into mony."

The key to stanching the plague of false coining, Chaloner asserted, lay in denying coiners access to the tools of their illicit trade. The difficulty was that everything used to make money also had a variety of legitimate applications, and cautious crooks hid behind the screen of honest labor. "It being lawful for them to keep such Tooles, in the Night, and other convenient times, they Coyn and afterwards break the Moulds." If they were any good at their trade, he said, in a brazen instance of self-revelation, "the mony being good Silver, it is difficult to discover them."

Chaloner proposed that a seal be placed on all instruments that could be used in coining. Only those who had "a Certificate from the Keeper of the said Seal" could "keep, sell, or dispose of any Sheers, Flatting-Mills, or Flasks." To obtain such a seal would require each applicant to bring with him testimony from "two of the Masters of the Parish they then live in . . . that they are of such Trades as do necessarily use such Tools in their lawful Employments."

In keeping with the conventions of the genre, Chaloner

carefully listed several objections to his scheme and responded with seemingly unassailable counterarguments. Some might say that it would cost metalworkers too much to comply. Not so, he responded, for even a busy goldsmith could not use more than two pairs of shears, for example. Perhaps coiners could use legitimate metalworkers as frontmen to buy "Sheers, Flasks, &c for them." But no! It would be easy to keep records of the buyers, and "if they offer to Buy more than two or three pair in seven Years, they shall be questioned, and suspected to be Coy[n]ers."

Most important, Chaloner explained, the whole idea of a black market in unlicensed tools was a fantasy. In all England, there were no more than twelve to fourteen master craftsmen capable of making the sophisticated metal tools required for a large-scale coining operation, most of them in London, no more than four in Birmingham and Sheffield. So small a group could be easily watched.

Even if Chaloner's numbers were off—and they probably were—they caught the underlying rhythm of England's transformation from a backwater into a true power in the world. There lived in the nation masters of the most sophisticated technical tasks then known—but not that many. Here was the daily reality in which Chaloner and Newton lived: a kingdom that traded in goods and knowledge across the globe—and that made its nails by hand.

Nothing came of Chaloner's suggestions, in the sense that Parliament ignored his advice. No act was passed to register metalworking tools; no effort was made to police instrument makers' workshops; no ledgers tallied how many pairs of scissors goldsmiths bought and sold. No matter: Chaloner was playing a long game, and for that his writings had the desired effect. *Proposals Humbly Offered* seems to have caught the eye of at least one important man: Charles Mordaunt, Earl of Monmouth and a former Lord of the Treasury, whom Locke had earlier approached as one of Newton's potential benefactors.

Chaloner's claim of knowledge superior to the Mint's made him potentially valuable in Mordaunt's own risky political gamesmanship. Once King William's confidant, Mordaunt had fallen out of royal favor by the early 1690s. Seeking a return to power, he was looking for possible weaknesses in his successors at the Treasury. His primary target was the man who had become Newton's patron, Charles Montague, Earl of Halifax, now Chancellor of the Exchequer. These two magnates had a long history of intertwined alliance and enmity. Their followers did not, yet. For the moment, each brushed past the other, unknowing.

In the meantime, Chaloner could not have been happier. Mordaunt's support finally forced the government to pay out the thousand-pound reward he had won for betraying the Jacobite printers two years before. Late in the year, with Mordaunt's encouragement and influence, perhaps at his order, Chaloner delivered a scathing indictment of the Mint's incompetence to control—or worse, its complicity in—the debasement of the coinage in testimony before the Privy Council.

It must have been a remarkable moment, the former runaway apprentice and sexual-novelty salesman entering the council room designed by Sir Christopher Wren in the palace of Whitehall. He came to speak to those who spoke to the King. If he could demonstrate that he truly understood the mechanics of moneymaking, and further, could persuade them that his expert knowledge had ferreted out villainy at the heart of the English monetary system, the ultimate prize might already be in reach: entry into the Mint itself.

It did not work out as he might have hoped. In that first appearance Chaloner could not quite convince his hearers that he was the man to set the Mint to rights. But his testimony was taken seriously enough to spark an investigation and to compel the Mint officials to respond to his charges. It was a good start. But before he could produce a more detailed account of the alleged corruption, he needed hard cash. And so, determined to "live as well as any person of Quality in the Kingdom by his Art, which he would follow," he declared, "spight of the Law," Chaloner now came up with perhaps the most inspired scheme of his fertile career.

Here is where Chaloner saw his chance. In August 1694, the Bank of England had opened its doors. Chartered specifically to raise capital from London's rich to lend to the government, it did something else as well, a kind of business never before seen in England. Day after day, clerks passed slips of paper—nicely decorated, to be sure—with numbers written on them, rather large numbers, in fact, and passed them to their customers. Those customers, wealthy men all, placed those papers in their purses or their pockets and walked out into London. They handed their papers on to others, men to whom they owed money—a tax collector at the Treasury, or perhaps a partner in some new business. Eventually that paper made its way back to the Bank. There, on demand, a clerk would fetch the appropriate stack of golden guineas or silver crowns and trade metal for paper.

To some—to Chaloner, certainly—the sudden appearance of what came to be called bank notes must have seemed like a gift from heaven: here at last was the road to wealth, paved not with gold but with sheets of England's first paper money.

13

"His Old Trick"

To most people in the 1690s, paper money was an oxymoron, as ridiculous and self-contradictory as a wise fool or a cowardly lion. Paper could not be real money. But faced with the cost of the war and the fact of a debased coinage, the demand for something—anything—that could pass between buyers and sellers, debtors and creditors, forced the issue.

The idea behind the Bank of England was hardly new. Prototype national banks had been tried in London in 1682 and 1683, and the key founder of the Bank of England, William Paterson, had offered his first proposal for a loan-making company to the government in 1691. But the idea of a central lending bank remained suspect, viewed as a way for investors to enrich themselves at the country's expense. Paterson's proposed bank would have charged the government six percent on loan capital of a million pounds, a rate of interest that was rejected out of hand by the House of Commons.

But by 1694, King William was desperate. The Treasury had tried to raise its own loan in 1692, and was forced to offer first ten percent, and then a ruinous fourteen percent, to attract less than 900,000 pounds—well under half the total needed to support the army in the Low Countries, never mind the government's other expenditures. When Paterson renewed his offer in 1694, this time for a bank with a capitalization of 1.2 million pounds, the Chancellor of the Exchequer rammed it through a still hostile Parliament during an almost unpopulated session of the House of Commons—reports indicate that only forty-two members showed up to vote.

In its final form, the Bank was supposed to provide a very simple service. Wealthy men would subscribe money to supply the Bank's capital, and the Bank would then lend that money—and only that money—to the government. Depositors could get access to their funds in three ways. They could hold a "book or paper" in which their transactions were entered—the passbook in embryo. They could write their own promises of funds to be paid, which the Bank would accept up to the amount of their deposits—the prototype of the check. Or, crucially, they could hold their money as "running cash notes," which the Bank promised to accept on demand, redeeming the note, either in full or in part, in hard coin. (Clerks would list partial payments on the note itself.)

That's how it began: money, captured on a sheet of paper. It rapidly became something more. By lending the full sum of its deposits (and soon enough, even more) to the government, and by issuing notes against that same capitalization that depositors could spend, the Bank of England performed the essential economic miracle: it created capital out of thin air. This was the birth of what is known as fractional reserve banking, the foundation of the modern financial system. In a fractional reserve bank, working on the assumption that only a small percentage of depositors will demand their share back at any given time, the institution lends more than the sum total of its capital. How much more is the question. Banks that lend too great a multiple of their deposits risk running out of cash if too many depositors demand payment. If the banking system as a whole lends too little, credit tightens, loans become more costly, and economic life suffers. (Bank regulators can use the reserve requirement—how much cash as a percentage of loans a bank is required to keep on hand—as a tool to tighten or loosen credit, and thus, in theory, keep an economy from becoming either too sullen or too exuberant. The gap between that theory and actual practice is not, perhaps, as small as economists would wish.)

At the start, the Bank had no vision of global capitalism. It was just trying to get negotiable credit into the hands of William and his army in Flanders — all while making a profit for itself. The unintended but critical consequence of creating modern bank notes, however, was that for the first time, a nation in Europe had transformed its government obligations into a new, uniform — and most important, tradable — commodity. Previous attempts to raise cash for the war had relied on any expedient available — loans, annuities, exotic proto-bonds — but none of these had an agreed value that enabled holders of an annuity, for example, to trade that asset on a market for cash. Bank notes *were* a form of cash.

The uniform character of the notes also meant that, useful as bank notes might be, they carried a potentially ruinous risk: what one man could inscribe on paper, another could copy. The raw materials for paper money were easy to find, and with so many printers and engravers in London, surely some would sacrifice their honesty for a price. After all, William Chaloner had persuaded a few of them to risk the noose by printing Jacobite propaganda.

The Bank's owners understood the danger, and they did what they could to protect themselves. The first bank notes were never intended to be ready cash, replacements for the silver and gold coins that ordinary people used to do their business. The new currency was supposed to remain in the hands of financiers who needed to move major sums of money around. Although the Bank offered notes in amounts as little as five pounds, one hundred pounds — or roughly double a middle-class income — was the most common denomination. Such large sums made it harder to pass poorly copied bills. Few would or could accept them, and those who did were mostly sophisticated enough to protect themselves from amateur criminals. But greater sums of money also increased the temptation. So, two weeks after the Bank received its charter, the directors formally decided that "the Notes for Running Cash being considered liable to be counterfeited, for preventing thereof it was ordered that they be done on marbled paper Indented."

Thus decorated, Bank of England notes—in practical terms, the first bank-issued paper money in the world—entered circulation in June 1695. They were immediately popular. As early as 1697, almost 700,000 pounds circulated as running notes—and this pile of what passed as cash quickly took on a life of its own. The five pounds that Mr. Smith deposited on Tuesday became ten by Wednesday: the five lent to support the army in Flanders and the five Smith could hold as a running cash note. This simple trick was the first in the sequence of novel gyrations of money that was about to turn London into the financial center of Europe and, within a century or a little more, of the entire world.

For Chaloner, marbled paper did not present much of an obstacle. He knew at least one artisan capable of simulating it, and he or a confederate was skilled enough to forge the handwritten entries on each note. His fakes circulated for at least two months before the first was detected, on August 14, 1695. If that was not enough time to match the haul of his early coining triumphs, with the house in Knightsbridge and his service of plate, it was still long enough to pass a serious amount of money.

But then, of course, he had a serious problem. The Bank's men traced that first detected dud note back to the printer who had copied the marbling. The printer informed on Chaloner, who then performed a magnificent two-step. Of course, he had to surrender his as yet uncirculated stock of false bank notes, but he forestalled prosecution with his usual expedient of trading information for freedom. He told the Bank's investigators of another scheme, one he had almost certainly also masterminded. The Bank had been fooled into accepting stolen checks from the City of London's Orphans' Fund, a fraud that had cost it at least a thousand pounds. Chaloner named names, and the infamous John Gibbons, porter at Whitehall, arrested those unfortunate enough to have been of use in this Chaloner enterprise.

It was a vintage performance. Chaloner played the boundaries between official London and its criminal shadow city better than anyone else in his day. For his services to the Bank—a double robbery, admitted and betrayed only after the profits had been booked—Chaloner received grateful thanks and, unbelievably, a reward of a further two hundred pounds. Encouraged—and why not?—Chaloner kept playing what must have seemed like an exceptionally gullible and rich mark, extending "his old Trick of Trumping up his Services." In November 1695, he sent the Bank a list of suggestions for dealing with the threat of counterfeit bank paper. His ideas impressed his readers. Sir John Houblon, the governor of the Bank, became an active supporter, to the point of seeking Chaloner's release from Newgate on his next brush with the turnkeys.

It was that same November that King William finally commanded Parliament to act in the face of the ever deeper crisis of England's coinage. To Chaloner, the decisions made in response to the King's demands only widened the field of opportunities. His experience in stealing from and then "funning" the Bank had confirmed for him that any turmoil in the money system created opportunities for him to get rich. Sixteen ninety-five had turned out to be a highly profitable year. It looked as if 1696 would be even better.

14

"A Thing Impossible"

Four a.m., late fall 1696.

The gate at the southwest corner of the Tower of London opens. Men appear from the darkness, some lagging, struggling against weariness and the aftereffects of the gin and ale consumed several hours before. A few of the bolder or more belligerent laborers chaff the sentries as they pass beneath the heavy stone of the gateway and through the Byward Tower. Scuffles break out between the soldiers and some of the workmen, and both sides wait for fighting words, or for the elbow in the gut or the flick of a leg that sends a man sprawling.

Past the tunnel that pierces the outer wall of the Tower, the stream of workers turns left. Another gate opens. Men enter the workshops that snake around the perimeter of the Tower—long, narrow rooms, airless and dark, propped up here and there by huge wooden posts, filled with rows of machines. As the hours pass, the air thickens—coal fumes and the stink of horse droppings merging with the funk of too many bodies working too hard in the rising warmth of the day. The noise is steady, relentless, rhythmic; the top note strikes fifty, sometimes fifty-five times a minute. Men push on through the clamor, the sweat, and the stink until early afternoon, when the next shift shows up. The Royal Mint, in full spate of what became known as the Great Recoinage—an enterprise unprecedented in the history of money—falls silent only at midnight. Four hours later, every day but Sunday, it lurches into action again.

This was how the recoinage ultimately proceeded—quickly, efficiently, relentlessly. But at the point when Isaac Newton took up his post as Warden, the effort—and at least potentially, the nation too—was on the verge of collapse.

The looming disaster derived from the fact that the man who was supposed to be in charge of the recoinage was thoroughly incompetent. In 1696 the Royal Mint was still an essentially feudal institution, headed not by one man but by three separate officers, the Warden, the Controller, and the Master and Worker. Each incumbent held his post by warrant of the Crown. There was no clear hierarchy of authority among them, and each had clearly fenced-off powers and duties. The Warden was nominally responsible for the Mint facilities alone. The production of new coins was supposed to be under the control of the Master and Worker.

Unfortunately for England, that meant that the fate of the nation's money supply in the spring of 1696 rested with Thomas Neale. Neale was a gambler's gambler, having served three kings—Charles II, James II, and William III—as groom porter, whose duty it was to furnish the royal residences with tables, cards, and dice, and to settle disputes among players. Neale himself played, on a grand scale. He asked for and received the first concession to create a North American postal service, a privilege for which he paid eighty cents a year. He hired a local deputy and lost heavily—three thousand dollars in the first five years of the service. He bet and lost another fortune on an expedition to recover a cargo of silver rumored to be worth more than one million pounds from the Spanish galleon *Nuestra Señora de la Concepción*, which had sunk north of Hispaniola. He wangled the Master's job in the usual way, through personal connections. But even with the benefit of patronage, his fame as a wastrel was such that he had to post a bond of fifteen thousand pounds of his own money, instead of the usual two thousand. Reckless, possibly corrupt, and certainly indolent, Neale was thoroughly overmatched by the job. When he died, it took his successor as Master—Newton himself—four years to untangle his official accounts.

In the meantime, Neale proved to be a true Mint traditionalist. His post had long been a patronage plum, and nothing could make him perform beyond what he believed were purely nominal duties. By 1696, he had long since turned over the bulk of his responsibilities to a hired assistant, with whom he shared a cut of the Master's profits from the coining operation. Such sloth did not matter during the slow years of the early 1690s. But with the recoinage, Neale was suddenly in charge of an operation that aimed, in three years or less, to melt and restrike almost seven million pounds sterling—more than the Mint had produced in the preceding three decades. No one in authority over the Master would have had any great confidence in his performance. But given that he held the royal warrant for his post, there was no obvious solution to his presence but to hope that the assistants he had already hired would be able to make up for their chief's deficiencies.

They could not. Under Neale's management, the first months of the recoinage swerved hard toward farce. The first significant milestone came in May 1696, when the Treasury ceased to accept the old, hammered money as legal payment for taxes. Neale's men were supposed to have produced enough new silver coins in the preceding five months to keep at least a reasonable fraction of usable silver in common circulation. But in the event, from May through July, there was almost no cash to be had anywhere in England, and the situation improved only slightly into autumn. Official revenues disappeared—tax payments abruptly dried up and government debt traded at a discount of thirty percent of its face value, a further drop from the already disastrous level of the year before.

It seemed as if the worst was about to happen. The mood in the nation shifted from complaint to near panic. Edmund

Bohun, formerly the Licensor of the Press—the official censor—wrote to a friend, "No trade is managed but by trust. Our tenants can pay no rent. Our corn factors can pay nothing for what they have had and will trade no more, so that all is at stand." Bohun expressed the general sense of terror. "The people are discontented to the utmost; many self murders happen in small families for want." Worse, he warned, "Should the least accident put the mob in motion no man can tell where it would end."

In June, the prolific scholar John Evelyn, a generally calm and well-connected observer, noted similar worries in his diary. There was a "want of current money to carry on the smallest concerns, even for daily provisions in the markets." It wasn't just small change that was lacking; the government was broke too. Between the cost of the war (the larger share of the problem) and the decline in the value of tax takings with the debasing of the currency, the treasure of the nation was exhausted. In Plymouth, an attempt to pay the army in old, worn coin raised the risk of mutiny—to the point where soldiers were paid in provisions instead of cash. Evelyn reached the same conclusion as Bohun: "Tumults are every day feared, nobody paying or receiving money."

In the town of Kendal, twenty people were arrested for rioting when a tax collector refused payments of the old, clipped money. In London, broadsides began to circulate blaming King William for the disaster: "Our Coine alas it Will not Pass / Which Makes ye World to Wonder"—wonder, that is, who was to blame. There was an easy answer: "Some say ye King contrived this Thing / his duchmen ffor to Cherish." Just in case anyone missed the point, the poet added: "In James time we had store of Coine / provision it was plenty." To all appearances, England had finally, literally, run out of money.

Newton entered the service of the Mint on May 2, 1696, having sworn that he would never "reveal or discover to any person or persons whatsoever the new Invention of Rounding the money & making the edges of them. . . . So help you God." Thus bound, his responsibilities included overseeing the maintenance of the buildings and machines and the care and feeding of the Mint's horses. But no one actually expected Isaac Newton to concern himself with fodder for the livestock or fixing broken windows, any more than previous incumbents had. That was what his three clerks were for. No Warden had done much real work for his four hundred and fifteen pounds a year for at least a century. (And after Newton, none would again until the office was abolished, more than a century later.) At any ordinary time, it would have been safe for Newton to take Montague at his word when the Chancellor assured him that the post would not be too demanding.

It took Newton no more than a few weeks to discover that this was no ordinary time. He measured the incumbent Master's qualities, coming to the common opinion that Neale was "a Gentleman who was in debt & of a prodigal temper." He took offense when he realized that Neale had more authority than he in the divided command of the Mint, and it galled him that such a wastrel made far more money than he did. Newton handled that problem by direct action—within a month of his arrival at the Mint, he asked for (and ultimately received) a raise to match the salary of the Master.

The larger issue of power in the Mint took a little longer to resolve. He immersed himself in paper and in meetings, making the decisions no one else could or would. In part, he was outraged at the level of neglect he found there. On May 6, just four days into his tenure, he sent the Treasury the painfully respectful suggestion that it might be a good idea to check the quality of the work of carpenters and laborers before paying their bills. In the next month, Newton wrote again, to complain that the Treasury had not supplied him with a budget sufficient to let him hire several needed employees. Sometimes he simply quibbled. Later that first summer, after tallying his expenses, he reminded the Treasury of a dispute over the grand sum of two pence.

More to the point, Newton quickly committed himself to mastering the details of every operation taking place in the Mint, including those that properly belonged in the Master's purview. He read up on the history of the Mint, tracing the records back more than two hundred years. He meticulously worked through decades of account books, annotating them in his own hand. He brought the rigor instilled by decades of painstaking laboratory work to bear on every step taken to turn raw metal into legal tender. He got his hands dirty as a matter of principle. As he told his deputies, his rule was to trust no other man's calculations, "nor any other eyes than your own." And through it all, he wrote. His holograph Mint papers fill five large portfolios, thousands of pages, a torrent of words.

As the summer of 1696 progressed, Newton's mass of knowledge accumulated into a weapon strong enough to bludgeon Neale aside. Faced with such overwhelming force, the Master had no hope, and he knew it. He surrendered, mostly quietly. He held on to that part of his pay he hadn't pledged against his debts, and left Newton to do his work for him. Apparently no one questioned this bloodless coup, although Newton lacked any official sanction to assume any authority over the Master's precincts at the Mint.

Newton now faced the same numbers that must have daunted the more experienced Mint officers. The coining machines had been designed to produce a maximum of 15,000 pounds of coins per week. At that rate, manufacturing the 7,000,000 pounds required to replace the entire silver coinage would have taken almost nine years. The Treasury ordered the Mint to ramp up production to between 30,000 and 40,000 pounds a week. But that, as Hopton Haynes, then a clerk assisting in the recoinage, reported, "was looked upon as a thing impossible."

By the end of the summer the impossible had become routine. Haynes, who became one of Newton's most trusted associates at the Mint, later said that the new Warden's cleverness with numbers (something of an understatement) enabled him to master the Mint's complex bookkeeping system faster than other men. That was certainly true. Newton was able to save the Mint from regular attempts to fleece it—like the time a pair of prominent metal dealers offered to take over the recoinage for the modest fee of twelve and three-eighths pence for every pound weight of silver to be coined. Newton ran a quick tally of the Mint's costs and demonstrated that these two benefactors were actually offering to overcharge the government by about a third. But it was the Warden's empirical skill—his ability to observe, measure, and act on his data—rather than his superior computational abilities that made the difference.

His first goal was to ensure that the Mint had the physical capacity to handle the recoinage. Mint workers crammed first one new furnace into the smelting room, then another. Newton oversaw the construction of a second melting house at the eastern end of the Tower walls. With all three of the main furnaces on line, the Mint could produce up to five tons of refined liquid silver suitable for currency each day.

That mass of molten metal flowed onto a decayed version of the assembly line that had so astounded Samuel Pepys. Now half a century old, many of the machines were falling apart, and those that still functioned were too few to handle the river of incoming silver. In response, at Newton's order, the Mint added eight new rolling mills and five new coining presses.

Next, the new Warden analyzed the potential performance of each stage of the coining process. He carefully observed the smelting operation, finding that each furnace consumed twenty-five bushels of coal per day. As in his alchemy, he made sure he understood the detailed characteristics of his instruments—for example, performing the measurement that revealed that a melting pot "which when new holds 800 lb wt, when it has been used a month or six weeks will hold but 700 or 650 lb wt or perhaps less."

Newton deployed the same empiricism on his men as on his machines. At the height of the recoinage, in late 1696 and through 1697, Newton commanded about five hundred men and fifty horses to drive the giant rolling mills. To ensure that this army wasted none of its efforts, he conducted perhaps the first time-and-motion study on record. As he observed, it took "Two [rolling] Mills with four Millers, 12 horses two Horskeepers, 3 Cutters, 2Flatters 8 Sizers one Nealer, thre Blanchers, [and] two Markers" to move enough silver from the melting rooms all the way down the line to feed two coining presses. Each press consumed seven more men—six to turn the capstan while one brave worker fed blanks into the striking chamber itself.

Those men constrained Newton's calculations. The Mint could not operate any faster than they could spin the capstan arms, and every other step had to be tuned to enable them to keep pounding out coins at the highest speed that human muscle and a driving screw could sustain. So Newton watched them work, to "judge of the workmen's diligence." He timed how long it took to strike each coin. He saw how quickly the brutal effort needed to turn the press wore out each team. He noted how nimble the man loading blanks and extracting finished coins from the press had to be if he was to keep his fingers. Eventually, Newton identified the perfect pace: if the press thumped just slightly slower than the human heart, beating fifty to fifty-five times a minute, men and machines could stamp out coins for hours at a time. That pounding set the rhythm that Newton used to drive the entire Mint.

Newton's drumbeat got results, fast. The record of the recoinage as a whole is one of an enormously complicated and expensive undertaking that was completed smoothly, efficiently, and mostly safely. (Only one man died at the rolling mills, an amazingly low number given the intensity of the work.) Under Newton's control, where once the sum of 15,000 pounds per week had been thought unattainable, soon the presses were turning out 50,000 pounds a week. By late summer of 1696, the Mint's men and machines achieved a record output of 100,000 pounds in six days—an unprecedented number, not just for the English Mint, but for all Europe.

At that rate, the recoinage raced well ahead of its original schedule. Most of the available silver was struck into new money by the end of 1697, and the entire project was essentially completed by the middle of 1698. In June 1699, matters had so far returned to normal that the Mint sold off the machines it had added to handle the national crisis. By then, the Mint under Newton's direction had totally remade England's stock of silver money, a total of 6,840,719 pounds. The total cost of the effort was huge—about 2,700,000 pounds, most of which represented the lost metal in clipped coins accepted for recoinage at face value. But for that price England had bought a whole new silver coinage with which to buy, trade, and fight.

The swift and ample transfer of silver coins from the Tower into public hands, beginning in the autumn of 1696, quelled the deepest fears of the day. There were no currency riots. The poor of London did not rise up to demand the return of good King James. King William continued to complain about the lack of money, but he was able to keep his army in the field, and by September 1697, after it was clear that the recoinage would be completed satisfactorily, he even achieved a peace with Louis

XIV. Nothing directly links the success of the effort with England's domestic calm or its military success abroad. But the fears that had seemed almost overpowering less than two years before disappeared from the record of public concern as the recoinage wended its way to a quiet, competent end.

Everyone knew who deserved the credit. At the conclusion of the recoinage, Charles Montague said that the enterprise would have failed without the presence of Isaac Newton at the Mint.

Part V

---❀---

Skirmishes

15

"The Warden of the Mint Is a Rogue"

FOR ALL THE PRAISE, honor, and wealth Newton's performance at the Mint earned him, there was one aspect of the Warden's work about which no one seems to have warned him before he accepted the post. By ancient practice, the Warden served as the Mint's only official magistrate, responsible for enforcing the King's law in and around London for all crimes committed against the currency.

Newton had no interest in the task, and he did his best to shirk it during his first summer at the Mint. He complained bitterly about the work to his superiors at the Treasury: "I am exposed to the calumnies of as many Coyners & Newgate Sollictors as I examine." A newly instituted reward granted forty pounds for each conviction of a coiner, along with a possible share in the convicted counterfeiter's confiscated property. Juries understood what such incentives could evoke and had become "so averse from believing witnesses," Newton noted, "that my agents and Witnesses are discouraged & tired out . . . by the reproach of prosecuting and swearing for money." Simply asking him to do the work at all was unfair: "I do not find that prosecuting of Coyners was imposed upon any of my Predecessors." Hence, he concluded, "I humbly pray that this duty may not bee annexed to the Office of the Warden of his Majts. Mint."

His prayers were denied. On July 30, 1696, the Treasury gave him the bad news. There would be no escape from his duty, and he was to start right then, with the vexing case of the disappearance of a set of coining dies from inside the Mint.

Nothing in Newton's prior career would seem to have prepared him for the sheer muddle of a criminal inquiry. Curves had properties that could be analyzed and relationships that could be proved. The behavior of bodies in motion could be observed and mapped against mathematical predictions. Theological argument could return to ancient texts, and rested always on the truth that God existed and acted in the world. To be sure, no one knew better than Newton how to shape a chain of cause and effect until only one possible conclusion remained. But here, there was no reliable measure with which to penetrate a maze of conflicting, chaotically human accounts. But he had no choice: the new Warden had to turn himself into a detective able to penetrate such confusion.

Newton's law enforcement career began with a simple question: what, in fact, had happened to the tools from the Mint?

No one could quite say.

It was possible to see how the affair began—or rather, first became known to the authorities. One day early in the year, Charles Montague, Chancellor of the Exchequer, found in his office a petition to the King and his Privy Council, dated January 13, 1696, and signed by a suspected felon confined at Newgate Jail, William Chaloner. Chaloner blamed his current predicament on the testimony he had given to the councilors the previous summer of wild abuses at the Mint. Mint officers had responded to the accusations by arranging with a private thief-taker to round up some of their usual coining suspects to testify against Chaloner, and had managed to get him committed to the cells pending the proper stitching up of the case that should have put a full stop to his career.

Despite this history, Montague does not appear to have fully grasped the significance of the signature on the petition he held in his hand. He may have recalled that someone by that name had been rewarded for his role in the discovery of Jacobite printers back in 1693. He probably did not remember, if he ever knew, that Chaloner had been held for coining at least once before, winning escape before trial when his accuser was put to death. Even if Montague could call that episode to mind, Chaloner's petition contained such a shocking, eminently plausible description of a conspiracy at the Royal Mint that Montague could not ignore the document. With the Great Recoinage just beginning, any hint of scandal could destroy whatever remained of public faith in the Treasury, so the Chancellor had no choice but to order an immediate investigation into Chaloner's claims.

Chaloner was duly released from Newgate. He returned to Whitehall on May 16, 1696. There, an investigative committee of the Lords Justices of Appeal heard him tell a harrowing story of official corruption and greed. In this sequel to his testimony of the year before, he repeated the claims of his petition: mint moneyers, the men entrusted to make true coins for England, were instead committing crime after crime. Using smuggled blanks made of base metal, they were producing counterfeit guineas on their own. When they did use properly pure silver or gold, they cheated the Mint and the nation by short-weighting the coinage. Worst of all, Chaloner testified, it was the Mint's own chief engraver who had sold off the official dies—the tools that struck the design into the faces of new coins—to coiners beyond the Tower's walls. Chaloner named names and swore that "he himself never made a Guinea in his Life," but he listed both his old confederate Patrick Coffee and, with marvelous bravado, a Mr. Chandler—a name known in the right sort of circles as the coining pseudonym of William Chaloner.

That was one story—horrifying enough to account for Montague's urgent reaction to the original letter. But was it true? Confounding the investigation, the tale was challenged on the spot by one Peter Cooke, described in his arrest record as "a gentleman," although he was already known to the authorities and presently residing in Newgate, where he was struggling to escape the death penalty in an unrelated counterfeiting case. With that incentive, he needed to be as persuasive as possible in his testimony, and the story he told the Lords Justices certainly captured their attention. Cooke admitted that he knew about the missing dies. But, he swore, those dies had not been corruptly sold out of the Mint. Rather, they had been stolen in a theft organized by a gang that included Chaloner himself.

Two incompatible accounts were bad enough. But then the Lords Justices heard from Thomas White, no gentleman, but like Cooke a convicted counterfeiter, testifying in the shadow of the gallows. According to White, the Mint itself and at least some of those who worked there had indeed conspired in what was growing into a massive counterfeiting scheme. White named a specific employee, a moneyer's man named Hunter, as the source of official dies sold to coiners. So far, it was a clear, straightforward story—until White added that Hunter had sold one set of dies to William Chaloner.

The swamp into which the inquiry had wandered grew soggier still when a Mint engraver known as Scotch Robin appeared before the committee. Robin corroborated Cooke's claim that the dies had been stolen, not sold. But the culprit he implicated was not Chaloner but Chaloner's accuser, Thomas White. When Robin himself came under suspicion, he ran, making his way to Scotland, safely beyond the reach of English writs.

Here the investigators seem to have given up. In this tangle of conflicting stories, only one fact could be stated with any certainty: someone, somehow, had gained illegal access to official coining apparatus. Beyond that, the mystery of the missing dies had become not so much a criminal conspiracy as a circular firing squad, with the growing army of the accused tumbling over themselves in their haste to betray one another.

Into the middle of this mess, under compulsion, came Isaac

Newton. He did not yet have any real knowledge about how to run a criminal investigation. He would prove an able student.

The jail at Newgate no longer exists. The earliest prison on that site admitted its first tenants in 1188. The last was demolished in 1904 to make room for an expansion of the Old Bailey. The jail in use in 1696 was almost brand new, constructed on top of the ruins left by the Great Fire of 1666. The façade of the re-built prison was given a hint of the elegance with which its architect, Sir Christopher Wren, hoped to endow the whole city. But such graces did nothing to alter the essential character of a place that was, as Daniel Defoe's Moll Flanders put it, not just "the emblem of hell itself" but "a kind of entrance to it" too. Defoe wrote from personal experience: he had been imprisoned there briefly, for debt. Other celebrated inmates confirmed Defoe's judgment. Casanova, imprisoned at Newgate under accusation of child rape, called it "this abode of misery and despair," an infernal place "such as Dante might have conceived."

This was useful terror, of course, and it began when a new prisoner first entered the underground holding cell beneath the main gate that the inmates called Limbo. Not coincidentally, condemned prisoners also waited there for their final ride to the place of execution, providing exemplary horror for the new arrivals.

There in the gloom, beside an open sewer cut into the floor, prisoners were taught the basics of life in Newgate. From that moment forward, bare survival—let alone any comfort—depended on how much cash the prisoner could feed his jailers. To be or become poor in Newgate invited disaster. New prisoners arrived bound in manacles and shackles on the hands and feet, neck collars for some. It cost two shillings sixpence "easement" to get rid of the ironmongery, and those who resisted could be persuaded. The jailers had been forced to give up the old technique of "pressing" inmates—crushing them beneath weights that were increased slowly, day by day—to get them to surrender their wealth. But there were still ways for an inventive or determined turnkey to encourage the miserly, such as tightening the metal collars enough to break a man's neck.

From the holding cell, prisoners were moved into the main prison. Richer inmates went to the Masters side. Those who could not pay the necessary bribes endured the Commons, where they were packed with up to thirty others into cells designed for a dozen or less. Beds were unknown in the common cells, so prisoners slept where they could, if they could. The diet was mostly bread, but an investigation in 1724 found that even those rations were routinely stolen by privileged prisoners—those who paid to handle the distribution of food and candles—some of which they sold to local shops. Starving, cold, condemned to rot in the dark, Newgate's most desperate residents went on suffering even when found not guilty of any crime. Prisoners had to pay a discharge fee before walking out the front gate, along with a charge for the food they might well never have received. No money, no exit.

Those in the Masters wards fared better. In what was called —and not in jest—the most expensive lodgings in London, prisoners with enough money could rent beds at three shillings sixpence a week, about a day's wage for a skilled worker. They could buy candles and coal, food and wine. The cells were less crowded, and the inmates organized them into something resembling a social order, with rank determined by time served.

Regardless of such relative comfort, the underlying fact of Newgate remained: it was a deadly place. Raw sewage; overcrowding; bad water for those who could not bribe their way to beer or wine; sleeplessness; cold and damp: put it all together and you had an almost purpose-designed incubator for disease. Typhus was so widespread there that mere remand to prison for any length of time could be a death sentence. Year after year, many more inmates died of the disease than lived to meet the hangman.

All this was what Peter Cooke and Thomas White, with their conflicting stories, faced from May to July 1696. Their executions were delayed and then delayed again, often for as brief a respite as a week or so, all to allow them to weigh exactly how horrible their lives were—and to consider how much worse (and brief) those lives could become. By the beginning of August, they finally achieved the proper frame of mind. Then, at last, the Warden of the Mint invited them to search their memories for any new information about the scandal at the Mint.

White faced the greatest and most imminent danger. The earlier case against him had been controversial, and its course illustrated the difficulty officials faced obtaining convictions for even notorious offenders. The evidence presented during his prosecution had been unconvincing at best. The Middlesex County grand jury had thrown out the charges against him three times before the prosecutor went jurisdiction-shopping and found a London grand jury that could be badgered into issuing an indictment. The persistence suggests that White had well-placed enemies, a hint confirmed after his conviction. A member of Parliament demanded his execution—and promised complications in the House if White was spared.

Newton's leverage was immense: if one conviction was not enough, by the time of their first interview, he had received information of another of White's crimes, information that White had helped two other men to set up a coining press—itself a capital offense. Newton was thus White's only hope, but at first it looked as if the convict had overplayed his very weak position. In his first interrogation he chose not to betray his two accomplices in the matter of the coining press, and Newton prepared to walk away, leaving White to the gallows. Just in time,

White began to talk. Newton held his canary tight, petitioning after each interview for a stay of execution of no more than two weeks. He kept the game going for months, delaying the hanging thirteen times in all, until he was sure that White had betrayed every person he plausibly could (and perhaps a few more besides). Finally, in May 1697, Newton let his singing bird out of the cage, arranging White's pardon after he had survived a full year in Newgate.

Peter Cooke grasped the essentials of the game much more quickly. He offered up at least three men as soon as he was asked. One was a deserter, and was promptly returned to the army. The second informed in his turn, productively enough to secure his own pardon. The third had nothing of sufficient value to offer. He was convicted and transported to the West Indies—the Fever Islands—a punishment that was understood to be a drawn-out sentence of death.

The information won Cooke his reprieve, but it did nothing to help Newton make sense of the case of the missing dies. Through August and September, he interrogated another six men—possibly more. He arrested more than thirty suspects, and as the autumn progressed, he set yet more investigations in motion.

What was William Chaloner doing while Newton and his growing crew of informers, runners, and clerks spread out across the city?

He remained in plain sight. After his release from Newgate, late in the winter of 1696–97, he found new lodgings in London. Newton had apparently interviewed him just once, possibly in August, certainly by the end of September. Everyone else caught up in the case had been questioned and threatened repeatedly by a relentless Newton until they broke. Chaloner alone stuck to his story that it was the Mint itself that harbored a criminal conspiracy—and Newton could not shake him from his claim.

As Newton was discovering, for all the severity of the law, coining proved to be a difficult crime to prosecute. Even securing an indictment was hardly a sure thing, as White's example demonstrated. Beyond the justified skepticism engendered by the reward system, the very bloodiness of the "bloody code"—the enormous list of offenses for which the punishment was death—made juries ever more reluctant to convict unless compelled by overwhelming evidence. In this case, William Chaloner had prudently kept himself at a safe remove from the incriminating dies. Cooke and White had been able to testify only that Chaloner had been somehow involved in the scheme, one of a group.

Thus insulated, Chaloner simply held firm, denying involvement, pressing his charges against the Mint, even offering to assist Newton in straightening out the scandal at the Tower. All Newton had to do, Chaloner told him, was to hire someone whom he could recommend without hesitation, Thomas Holloway—just coincidentally, Chaloner's old coining partner—to serve as supervisor at the Mint.

Newton brushed that aside. Though still in his first months as a crime fighter, he knew better than to accept "help" from suspects. But the fact remained that there was no obvious reason to hold on to Chaloner. He didn't have the dies. Those who accused him of involvement were already under sentence of death for separate matters—which meant that any jury could reasonably discount their testimony as an act of desperation. And Newton did not yet know whom he was really dealing with in William Chaloner.

In our age of constant communication, we must keep in mind how hard it could be to keep track of the bad guys in Newton's time. Chaloner had surely left enough tracks to register as a rogue. Newgate's keepers would have recognized him from his two previous visits. Some in the government would have remembered him from the affair of the Jacobite pamphlets, and someone at the Treasury should have recalled the thousand-pound reward paid him in 1695.

But England would not see a modern police force until Robert Peel established the world's first: the Metropolitan Police that began to operate in London in 1829. With the Met, bureaucratic notions of routine record keeping, the sort of dull, effective note taking and filing that would permit the police to keep track of its villains, finally became routine. But in 1696, Chaloner could bank on the fact that this kind of policing did not exist. Identification was haphazard, anecdotal. Those taking on police functions had as yet no obvious way to talk to each other in the usual course of events. Agents of the Crown chasing political conspiracy had no reason to flag evidence for the Warden's men seeking counterfeiters. Whitehall might be in possession of a rap sheet pages long for a given criminal, and no one at the Tower would have a clue.

Which meant that to Newton, in August 1696, Chaloner was just one more shady character in the huge and indiscriminate pile of information with which he had to work. Newton knew that witnesses talking in the shadow of the gallows would say anything they could to escape, so there was a cloud of doubt as to Chaloner's actual role in the crime at hand. Under the circumstances, no conviction was likely. Newton had no real option but the one he took: he asked Chaloner what questions he could. Chaloner answered, careful to avoid contradiction. Newton listened, and then let his man go.

From Chaloner's perspective that escape was a victory, even though he had failed at part of his scheme. He had not been able to entice the new Warden into taking Thomas Holloway into the Mint. Yet he had done well. He had accused the Mint of criminal conspiracy and gotten away with it. The whiff of scandal still hung over the Mint and its officers, and Chaloner had managed to fade from view.

For Newton's part, this first encounter with Chaloner did not register very deeply. He was a busy man. His real work, if not his official duty, lay with the ongoing recoinage. Until and unless Chaloner again chose to inject himself into Mint business, he was in no danger whatsoever from Isaac Newton.

Chaloner drew his own conclusion from the seeming ease with which he had wriggled out of jeopardy. Even though the Holloway scheme had failed, there were other ways to take advantage of the confusion in the country's money. By the spring of 1697, he hit upon a new plan that should have allowed him to dip at will into the river of wealth flowing through the Tower of London. He had, he believed, no reason to fear the inevitable result: a head-to-head confrontation with the apparently easily fooled Warden of His Majesty's Mint, that unworldly natural philosopher so recently come from the provinces.

16

———✦———

"Boxefulls of Informations
in His Own Handwriting"

FORCED TO BECOME a criminal investigator, Isaac Newton
committed himself to doing the job well. In August and September 1696, he devoted as many as half his working days to the
case of the missing dies. Once he got through the initial round
of interrogations, he paused to consider just how a proper investigation should be run.

He soon established his basic strategy. He knew that coining
was of necessity an organized crime. And if he could not work
that out from first principles, Cooke's and White's testimony
taught him the facts of the counterfeiter's life: it was impossible to coin on any scale without confederates—which meant
that there were always at least three or four people who could
bear witness against each other, even before a single dud guinea
reached the street.

That was where the greater vulnerability lay, of course; the
problem of making an illegal sale has plagued aspiring crime
lords throughout recorded history. Then as now, master counterfeiters did everything they could to avoid direct contact with
the street-level trade, selling off coins in large lots to buyers who
would then recruit others to pass the money into daily use. But
there was still a chain of contact that could trace a single false
piece back up the line. Worse, given the ratio of risk to reward
at the bottom of the food chain, street-level traders in bad coin
had every reason to talk when caught. In theory, and sometimes

in practice, even a trivial quantity of bad money could doom someone to a coiner's death. And always, while captured suspects waited to see what clemency might be theirs, the evils of Newgate remained an effective way to loosen tongues.

All this dictated Newton's approach. To break the threat of counterfeiting, he would need to capture and convict major players. To do so, he had to have witnesses and physical evidence that would link those players directly to the crime, to the making and disseminating of coin—a strong enough link so that even the most soft-hearted jury would convict. To find that evidence and to connect it to the men he wanted, he needed to roll up the networks no counterfeiter could do without, starting at the bottom, trading precisely measured grains of mercy for the knowledge he needed. Like any street cop in history—and unlike any other fellow of the Royal Society or Cambridge don—he would have to wade hip-deep into London's underworld.

He begun to do so no later than September 1696. While pursuing the implications of Cooke's and White's confessions, he recruited his first agents to develop other cases, men he sent out in an unprecedented sweep of undercover operations. On September 11, 1696, his accounts record five pounds "paid Humphrey Hall to buy him a suit to qualify him for conversing with a gang of coiners of note." That was some suit: five pounds was a month's pay for a clerk at the Mint. Newton was clearly aiming high here, sending Hall to mix with a flash crew, notable crooks who dressed to match their success.

Over the next several months, Newton ranged further in the hunt. To solve any problems of jurisdiction, he got himself appointed a justice of the peace for seven counties surrounding London's metropolitan sprawl. Thus armed, he dispatched men in pursuit of coining rings wherever the evidence led. An agent who lived in the London suburb of Islington traveled back to Newton's recent home, Cambridgeshire. There, posing as a coiner on the lam from the capital, he wormed his way into a fully equipped coining operation, complete with furnace, flattening mill, and a version of the Mint's "secret" milling machines.

These investigations were expensive. The brothers Benjamin and Charles Maris, who traveled through Worcestershire and Shropshire in late 1696, billed Newton £44 2s. for wages, expenses, and tongue looseners. Bodenham Rewse, also known as Benjamin Reuss, listed in court documents as an embroiderer living in Bow Street, actually made his living as a thief-taker. Between 1693 and 1695, he and a partner brought charges against twenty-two prostitutes and more than a dozen owners of brothels. But his career really took off when he entered the Warden's service. Newton clearly trusted him, giving him several arrest warrants to execute, and then paying him £34 to pursue coiners operating to the west of the city. Both men benefited. Newton was able to prosecute several coiners based on his man's investigations, and by 1701 Rewse had picked up enough spare change in bounties and rewards to be able to buy the post of head turnkey at Newgate—thus reaching the top of the system of more or less legal plunder that could turn jailers into rich men. In all, when Newton accounted for his out-of-pocket expenses incurred in the pursuit of coiners from 1696 to 1699, he came up with a total of £626 5s. 9d., or well over a year of his own salary as Warden. That was more than enough to support a serious private force, answering only to Newton, to be aimed at whomever he chose.

Inevitably, some agents turned bad. By 1697, both the Maris brothers ended up on the wrong side of the locks at Newgate, one for smuggling, the other for coining. Others the Warden employed were worse. Hopton Haynes, usually almost worshipfully loyal to his patron, acknowledged that the agents Newton recruited "lay under violent suspicion of being scandalously mercenary." There was Samuel Wilson, who confessed to Newton that he had sold "a pair of Dyes for making mill'd Shilings" for five pounds. Newton gave his informant a Warden's warrant to arrest the buyer—and Wilson seized on the document as a gift, "as good a Sham as could be to get money." He used the warrant to blackmail victims for a year and a half before he was betrayed in turn.

Then there was the terrifying John Gibbons, porter of Whitehall and already one of William Chaloner's more valuable contacts. The authorities in London, Newton among them, used Gibbons for years as a thief-taker, charged with performing what would become essential police functions: running informers, searching suspect premises, executing arrest warrants. Gibbons used these powers to operate a lucrative side business. He did arrest those who forced him into such unprofitable inconvenience, and he pocketed whatever bounties he could claim, but his protection racket earned him much more.

Newton eventually recognized that his man had gone too far to the bad. He turned his attention to the problem in the spring of 1698. Witness after witness told him how thoroughly Gibbons had come to terrorize the coining world. One informant told Newton that paying Gibbons "a certain pension quarterly and yearly" was a fact of life, part of any London coiner's overhead. Gibbons's onetime lover Mary Townsend testified that he had been running his protection scheme for at least six years, while a captured coiner named Edward Ivy (aka Ivey) confirmed that the racket was still going strong: "Gibbons corresponds with a great many Clippers and Coyners and used to receive severall summs of money from them as contribution for coniving at ym. [them] and was wont to Solicit for any of ym. when they were in restraint." Gibbons's standard price seemed to be fifty pounds, though he would occasionally suggest alternatives, and he did not always, or only, demand cash. Elizabeth Bond told Newton that she had seen Gibbons lead a Mrs. Jackson into "a little adjouning room with a Bedd" and that when they emerged, "Mrs. Jackson trembled with her hands and Jawes and lookt pale." Jackson could merely have been terrified, but the pointed mention of the bed suggests something more, and the hint of sexual extortion runs through the witnesses' accounts. At the height of his power, Gibbons put the bite on most if not all coiners who came to the attention of the authorities, among them William Chaloner, who, Gibbons boasted, was "sought after for coyning of Gineas and pistols" but was "safe enough for he had secured him."

Newton neither participated in nor tolerated his agents' excesses, but, then as now, corruption was an inevitable byproduct of policing any highly profitable illegal business. The fact that he had to use thieves, extortionists, and coiners themselves to capture those he wanted more did not much matter to him. Partly the problem was self-correcting: the worst of his men tended to overreach to the point where he could deal with them as needed—and in the meantime his thugs got results. By early 1697, Newton's network of informers, undercover agents, and street muscle had turned him into the most effective criminal investigator London had yet seen.

Newton was willing to stick his own oar in too, as needed. In October 1699, he submitted another bill to the Treasury. He asked for £120 to cover "various small expenses in coach-hire & at Taverns & Prisons & other places." He spent that sum on personal forays through London, buying drinks for informers, sweetening the deal for accomplices—all in all, diving as deep as needed into the muck of the capital's criminal landscape.

He did not shy from the more brutal side of the job either. He personally examined those he and his men caught—showing up in the cells at Newgate if necessary, hauling informants into the Mint's narrow, private chambers at the Tower if he could. Typically, Newton would ask the questions and take notes while a clerk wrote a summary of the deponents' confessions, to be signed once the ordeal ended. Most of those documents have disappeared, perhaps suspiciously. John Conduitt, Newton's nephew by marriage and his successor at the Mint, reported that he helped Newton burn "boxefulls of informations in his own handwriting."

Conduitt chose not to explain why Newton wanted to destroy the papers, but one inference is that Newton enjoyed the role of inquisitor too much. In this view, Newton proved willing, perhaps eager, to terrorize his captives in pursuit of the necessary confessions and betrayals with a viciousness that even that strong-stomached time would tolerate. Formally, torture had not been used in England as an investigative tool for about half a century before Newton came to the Mint. Elizabeth I had faced repeated rebellion, often animated by Catholic ambitions on her Protestant throne—and she was England's most prolific torturing monarch, authorizing fifty-three of the eighty-one warrants on record. The rack was the most common tool used to extract confessions, but occasionally Elizabethan inquisitors grew more inventive. On November 17, 1577, Thomas Sherwood was consigned to a dungeon overrun with rats, and on January 10, 1591, four torture commissioners were ordered to confine the dangerous priest George Beesley and a co-conspirator in the tiny cell known as Little Ease. There, Beesley could not sit or stand or move at all.

The last case of legal torture in England came in the spring of 1641, after a riot of about five hundred people at the Archbishop of Canterbury's palace at Lambeth. One man was identified and arrested: John Archer, a young glover or glover's apprentice. He had made one disastrous mistake. In the midst of the tumult, he took up a drum with which to urge the crowd forward. That turned a Monday night's rough party into the crime of sedition, for "marching to the beat of drum was held to be a levying of war against the king."

Archer did not name any ringleaders, so on May 21, King Charles I issued the last English torture warrant. It ordered the lieutenant of the Tower first to show Archer the rack. He looked; he remained mute, refusing to name names. That silence invoked the second part of the warrant: his interrogators were "to cause him to be racked as in their discretions shall be thought fit." At this point, the Tower's men had to place Archer on his back beneath the frame of the rack with his wrists and ankles bound to two rollers. Men at the levers attached to each roller heaved on order. Archer's body rose until it was level with the frame. With the next pull, the rack strained on his limbs, starting his bones from their sockets, threatening his fingers and toes, hands and feet. He may have been stoic beyond imagining, or he may have been just one more rowdy man who got mixed up in a crowd and truly had no one to betray. Either way, Archer did not speak. In the end, his tormentors gave up. He was hanged the next day.

But for all the pain and retributive satisfaction produced by torture—James I famously blessed the torturers who worked on the assassin Guy Fawkes, saying, "God speed your good work"—officially sanctioned torment fell out of favor, at least as a tool with which to acquire evidence fit for a criminal trial. England's use of juries helped dampen enthusiasm for the practice: juries could convict on any evidence they chose, and thus did not need a confession wrung out of an offender's howl of pain. There was even appreciation of the fact that confessions, or any evidence gained under torture, might not be entirely reliable.

But while official torture fell out of favor, interrogators still knew how to put the boot in as needed. Isaac Newton had plenty of ways to extract the information he wanted from reluctant prisoners, and he made use of all of them. Most of them were within the customary bounds of police detection: trading in fear, not pain. He offered brief reprieves for information; he coerced husbands with threats and promised rewards to wives and lovers. But there is one—and only one—reference to his use of more brutal methods in the records he did not burn. In March 1698, Newton received a letter from Newgate written by Thomas Carter, one of Chaloner's closest associates. The letter was one of a flurry of messages Carter had sent to confirm that he was eager to testify against his former co-conspirator, but this one had a postscript. "I shall have Irons put on me tomorrow," he wrote, "if yo[ur] Worship not order to the contrary." In other words: Don't hurt me! Please. I'll talk. I'm ready.

An ugly moment. Shackling in irons is not the same as racking someone, but the capacity to inflict terrifying pain was there. Some historians have simply condemned Newton, viewing what they see as the bloody ruthlessness of his pursuit of counterfeiters as evidence of a damaged mind, a heartless man. Frank Manuel, one of the most influential of Newton's biographers, argued that what he saw as Newton's pleasure in the chase and ultimate execution of coiners was a kind of catharsis of the anger and loss that had driven him mad in 1693. "There was an inexhaustible font of rage in the man," Manuel writes, "but he appears to have found some release from its burden in these tirades in the tower." He added, "At the Mint, [Newton] could hurt and kill without doing damage to his puritan conscience. The blood of coiners and clippers nourished him."

This is almost certainly nonsense. There is no record of Newton's gloating over his victims, or of his being present during any attempt to use physical coercion to extract information. Rather, he was a familiar figure, just doing his job, a bureaucrat using the means that were generally available at the time. Everyone involved in the criminal justice system had recourse to the ordinary miseries of imprisonment, privation, and then, as needed, the back room and all its terrors. The threat alone was probably good enough for most purposes. From the records that do survive, it seems that Newton, like most other English officials, did not torture in the legal sense (for all that, it also seems plausible that at least a few of those in his custody suffered some bodily harm). He did not have to. The same reasons that led sovereigns to discontinue the practice would have held good for Newton too.

Yet the essential fact remains: Newton, only months removed from the life of a Cambridge philosopher, managed incredibly swiftly to master every dirty job required of the seventeenth-century version of a big-city cop. He found in himself the capacity to do what had to be done.

17

"I Had Been Out Before Now but for Him"

MOST OF LONDON'S coiners did not grasp the danger this strange new Warden posed. The documents Newton did not burn, all written between 1698 and 1700, reveal the almost unfair contest between the Warden and those who tried to trade in bad money. One case in the Mint files tells of a conspiracy from the summer of 1698. Early in July, a man named Francis Ball came to the Crown and Sceptre in St. Andrew Street in the City of London. The tavern had a tough reputation and was known to be something of a clearinghouse for jobs the authorities might not approve of. Someone there told Ball about Mary Miller. At the time, Miller was down on her luck, but she was known to have passed bad money, most recently trading in shillings made of pewter. From Ball's perspective, she was the perfect accomplice, broke and able, and he told her that he and his friends "could put . . . [her] in a way whereby She might be serviceable to them and her Self and get Some money by it."

Ball's proposition: he had made or bought twenty false Spanish pistoles (two of Newton's witnesses differed here). Now he had to get rid of them. He asked Miller to take his stock to one of her contacts in the game. She tried to resist temptation—at least she claimed she did, in her own, probably self-serving account. She told Ball she lacked clothes respectable enough for the job; Ball said he'd buy her better ones. She told him that she knew only one person who might be interested; Ball told her he would take his chances. She got up and left the Crown and Sceptre twice. Each time Ball called her back. At last, she gave in.

Later that day, Miller took two of the counterfeit coins to a house in Smithfield, home to both the meat market and the execution ground where coiners had been taken to their deaths in the previous century. She showed her pistoles to a Mrs. Saker (aka Shaker), left one behind as security, and arranged a meeting with Ball at a nearby tavern the next day. Ball and Miller both came. Saker was waiting. Ball tried to put some distance between him and the crime by handing Miller the paper wallet holding the false coins. Miller dutifully passed it on to Saker—who turned out to be a plant. Her husband burst into the tavern with several other men, seized the false pistoles, and arrested Ball and Miller.

Ball, having made his first mistake by hiring Miller, now compounded the catastrophe by using her as his messenger. When it seemed as if she might be released, he asked her "to go to one Mr. Whitfield to desire him to stand . . . his bail." He told Miller that "the blanks for ye counterfeit Spanish pistoles were cast . . . in Whitfield's house." That made it urgent to get rid of their coining tools before "worse came on it." Here was evidence that could hang both men, but Miller put Ball off, telling him that it was too late that night for Whitfield to be able to do anything. In truth, it was. Mary Miller was a known quantity to Isaac Newton. She had already betrayed Whitfield—or at least a woman named as Whitfield's "particular friend" accused her of doing so. Within a day, perhaps two, both men found themselves sharing a crowded cell in Newgate.

There they sat, fuming. Ball and Whitfield remained inside for more than a month. By mid-August Ball had had enough, and he told his friend so. "Damne my blood," he said. "I had been out before now but for him," meaning Newton. Whitfield agreed. Their nemesis "was a Rogue and if ever King James came again he would Shoot him"—treason upon treason here, first for suborning the King's coin, and now for giving voice to the Jacobite dream of overthrowing King William. Ball happily cheered this double treachery: "God dam my blood so will I and tho I dont know him yet I'le find him out."

At least one man in that jammed cell was listening. Once Newton put his mind to the job, someone was always listening. This time it was a man named Bond, Samuel Bond.

Bond was a "chyrugeon," a surgeon; he was originally from Derby, now lodging in Glasshouse Yard in Blackfriars. He had been arrested for debt, and he had a fine memory for dialogue. His testimony completed the case against the counterfeiters. In addition to the threats against the Warden's person, Bond told Newton that he had heard the two men plan to renege on their bail, and had listened as they described how they gilded their false money; he reported that they said it cost them no more than six or seven pence per coin to do so.

This was how it was supposed to happen: coiners, traitors, condemning themselves out of their own mouths. Newton placed himself—or those he had persuaded or compelled to serve as his eyes, ears, or provocateurs—into position to catch them as they blabbed. Ball and Whitfield, puffed with bravado, completely overmatched, provided a proof, elegant as any geometrical demonstration, of the proposition that one tangled with Isaac Newton at one's desperate peril.

While talkative fools like Ball and Whitfield presented little difficulty to Newton, William Chaloner remained at large, a wholly different species of problem. He was vastly more ambitious than the usual optimists Newton confronted. "He scorn'd the petty Rogueries of Tricking single Men," aiming instead at "imposing up on a whole Kingdom"—as his biographer rather proudly put it. Although true-crime writers tend to inflate the importance of their subjects, Chaloner did in fact aim to play on the national stage—and unlike any other coiner Newton encountered, Chaloner had the wit to take a genuinely long view. The scheme he set in motion in the spring of 1697 had its start in the moves he had made over the previous three years to gain some authority over the Mint itself.

Chaloner, like Newton, understood that counterfeiting carried with it the certainty of betrayal. The coiner, forced to rely on others to get his product into circulation, knew that some of his confederates would be vulnerable to arrest and then to the need to save their own skins. Chaloner had already grasped that there was one sure way to get coins into the marketplace without openly passing them: do so from within the Mint.

He had tried to get inside that magic circle already, but Newton had not fallen for his trick. His next attempt to insert himself was better prepared. In February 1697 Chaloner appeared before a special committee of the House of Commons investigating alleged abuses at the Mint. He provided the committee with what seemed to be a compelling account of Mint errors and his own plausible remedies. He began by arguing that the Mint's officers were incapable of detecting counterfeiting, or were even party to a sophisticated version of clipping taking place right in front of them on the coin production line. Chaloner told Parliament's investigators and then argued in a brief pamphlet that the Mint's chief employees were so specialized that they could not check each other's work for fraud. "None of the said Officers of Work-men," he wrote, "know whether the Essay-Master hath made the Bullion standard," nor if "the Melter doth Mould and Temper the Bullion so fit [i.e., to make it suitable] for the Impression" (the stamping of coin faces), and so on, until, Chaloner declared, "Now everyone doing his business as may be most for his own advantage."

The result of such carefully maintained specialization? According to Chaloner, it had already been proved that "there hath been a great quantity of Counterfeit Mony Coyned in the Mint"; that Mint staff were selling dies out of the Tower; and that "our present Money is so disengenously Coyned that it may be easily Debased, Diminished and Counterfeited."

The worst of it was that at least some of what Chaloner alleged was true. Dies had been spirited out of the Mint. Counterfeiters were producing false coin. Individual Mint officers did execute their duties, as Chaloner put it, "as may be most for his own advantage." The rot started at the top, with the Master, Thomas Neale, who raked off his percentage of every piece struck during the recoinage—amounting to more than fourteen thousand pounds in 1697 alone—and yet did nothing for the money, delegating the actual work to a relatively poorly paid salaried assistant. Further down the chain, as the committee reported with scorn barely masked by the official prose, "the present assay master and the present melter have married two sisters."

Why should it matter that the two men had become brothers-in-law? Because, although the previous melter had given up his post, unable to make a profit at the agreed price of four pence per pound weight of silver committed to the furnaces, the current, happy husband "hath got a great estate by this place and keeps a coach." Here, unstated but clear, was the implication of corruption. There was only one way the new melter could gain such easy riches, when his predecessor could not. The assay master, his wife's sister's husband, had to be sending along silver mixed with excessive alloys of cheaper metal—a fraud that would allow the two men to pocket the difference.

This, clearly, was a disaster in the making. If it became generally believed that the Mint was releasing what were, in effect, clipped coins, the value of English currency would again become a fiction. Still, the particular form of the fraud suggested an obvious solution, as Chaloner eagerly pointed out. Given that the management of the conniving, conspiratorial, money-grubbing Mint workers had proved to be beyond the Mint's top leadership—no need to specify whether it was the absent Neale or the inexperienced Newton—why not add "an Officer . . . to the Mint, who understands Melting Essaying, Alloying Graving, Smith-work and all other parts of Coyning"? That man "shall supervise the Word and Assay the Money when Coyned," reporting under oath each month the results of his efforts.

No one needed to be told who this paragon should be. Chaloner knew, however, that merely diagnosing the faults in the management of the Mint would not win him this notional supervisor's post. And so, in a demonstration of what made him unique among London's counterfeiters, he proposed to the parliamentary investigators a test of his ability to handle the job. Chaloner told them he had invented a new technique for making coins, an approach he described as "A Method . . . humbly Proposed, how Money may be Coyned, so that it will be Morally impossible to Counterfeit."

All counterfeits, he reminded the committee, are made "either by Casting or Stamping." He had ideas about how to deal with both methods. To defeat those who could cast good simulations of the Mint's edged coins, he proposed a new technique and machine—one that would mill the edges of coins "with a Hollow, or Groove." Such a subtlety would render it "certainly impossible to Counterfeit Money by Casting it." To prove the point, Chaloner asked the committee to take one of the samples he had struck and pass it on to the guild of goldsmiths, to get their assurance that the new method could not be copied. This was classic Chaloner. He had never yet seen the inside of the Mint. He had, for all his efforts, no official role in managing the currency; he had been repeatedly implicated in coining on his own behalf over the previous five years. And yet now, with his own hand, he gave Parliament the proof that he could at will commit what was, after all, a capital crime.

Chaloner's next move brought him to the point of the entire exercise. Having proposed a technique to thwart those who cast their counterfeits, he offered a new plan to defeat those who stamped their products. Current coins, he said, were of "such bad Workmanship so that every Graver, Smith, Watchmaker &c. can Grave Stamps to counterfeit Money, and Stamp it with a Hammer, upon a Stone." The incompetents at the Mint could not outthink a mere blacksmith—but Chaloner could. He had brought with him the materials to show the worthy men of Parliament how the nation's currency ought to be produced. All it would take were a few minor alterations to the Mint's machines, which he could make in the moneying rooms at the Tower in a few days and for a modest charge—no more than a hundred pounds or so. Then the modified coining machines would be able to "Stamp . . . the Impression so high as to make it Impossible to do with a Hammer or a small Engine"—and even better, his improved methods would require only "two Horses . . . [to] do all the work, which now Imploys 70 or 80 Men." When all was in place—new tools, modified methods, and a couple of willing horses—then, Chaloner wrote, the new money would be "more Beautiful, and Durable than now our Coyne is made."

And what, besides a bit of cash and a week or so, would be required to achieve such a triumph? Not much, a mere trifle: just that "The Proposer hereof being order'd to perform some of his Proposals in the Mint." As ever, Chaloner kept his eye on the prize: access to the Mint, its tools, its river of hot, precious metal.

The implications of Chaloner's testimony were missed by no one—certainly not Isaac Newton. In a response to Parliament, he wrote that "Mr Chaloner before a Committee of the last session of Parliament labored to accuse and vilify the Mint." Newton's responses were consistently defensive—discordantly so, given his usual tone of absolute authority. In one draft memo he wrote, weakly, that he was not to blame for the behavior of the assay master and the melter because their false coining at the Mint happened "three weeks or a month before ye Warden knew any thing of these matters." He went on to complain that some of his own testimony had been omitted from the committee report, as if he had been reduced to mere procedural objections.

Nevertheless, Chaloner did not entirely convince the members of the committee, not even with his bravura demonstrations of what a coiner of real skill could do. But he did impress them. They found that "undeniable demonstrations have been given and shewn unto this committee by Mr. William Chaloner, that there is a better, securer and more effectual way, and with very little charge to his majesty, to prevent either casting or counterfeiting of the milled mony . . . than is now used in the present coinage." And so, on February 15, 1697, the committee commanded Newton to "prepare or Cause to be prepared such matters and things"—inside the Mint—"to the End [that] the said Mr Chaloner may make an Experiment . . . in relation to Guineas." That is, if the committee were to be obeyed, Isaac Newton had to welcome into the Mint a man who had just argued as publicly as possible that the Warden of the Mint was a fool, a thief, or both.

Newton chose not to comply. He had legal grounds to refuse the order. The oath he had sworn on taking up his post bound him never to allow an outsider to see the Mint's edging mills. Instead, he asked Chaloner to tell him how his methods worked, and when Chaloner refused, took it on himself to "direct the workmen (without him) to groove some half crowns, shillings and six pences." Newton himself carried those coins to the committee, demonstrating that Chaloner's ideas were unworkable. And there the matter rested, at least officially. If the House was offended at the Warden's recalcitrance, it did not stop its investigative committee from pasting a large section of Newton's testimony into the final report, verbatim.

But the fact of Chaloner's charges remained a public stain. Chaloner continued to press his claim through the spring of 1697, still hoping that the pressure of parliamentary patronage would win him entry to the Mint. It did not. He had miscalculated—though it was not yet obvious how badly. Newton had been perfectly ready to forget William Chaloner after the messy business of the missing Tower dies the year before. But the parliamentary report, with its praise for Chaloner, was an open sore. Through page after page of draft rebuttals, written in a cramped and crowded hand, passages crossed out and written over in tiny, hasty, furious script, great gobs of ink blotted here and there, runs Newton's private rage. He complained of "calumny" and of the offense given by Chaloner's "libeling . . . in print." Publicly, though, he held his tongue. He waited and he watched, he and his agents, eyes and ears open all across London.

18

"A New and Dangerous Way of Coining"

Two BRUSHES WITH Isaac Newton had done nothing to diminish Chaloner's sense of invulnerability. He continued to hold out hope that despite Newton's resistance, he would yet be granted a powerful post within the Tower's moneying rooms. He boasted to his brother-in-law that having "fun[ne]d the Lords of the Treasury and the King out of 100 pounds," he would not leave Parliament "till he had fun[ne]d them likewise."

Such confidence must have made what happened next a truly galling disappointment. Newton proved able to sustain his defiance of Parliament's order. Chaloner was not to be admitted to the Mint under any pretext. He could not use the Mint's machines to demonstrate his ideas. He would not be asked to join the Mint's staff in any role, much less that of supervisor. According to Chaloner's biographer, the investigating committee finally saw through the persuasive coiner: while he had "accus'd that Worthy Gentleman Isaac Newton Esq; Warden of this Majesties Mint, with several other officers thereof, as Connivers (at least) at many Abuses and Cheats," in the end, the committee "appointed to examine the same . . . upon a full hearing of the matter, dismissed . . . Chaloner with the Character he deserv'd."

It did not happen exactly that way; the need to preserve the appearance of a morality tale made the polite lie necessary. In fact, the endorsement of Chaloner's anti-counterfeiting measures in the committee's public report obscured the underlying political reality, which was that the committee was the work of Charles Mordaunt, Earl of Monmouth, and his friends, this time seeking the Master's post at the Mint for an ally. Newton, and even the feckless Thomas Neale, had seen the investigation for what it was: part of a larger, longer game of parliamentary maneuver. Both, however, were members in good standing of England's ruling faction, and they knew that there was never the slightest chance that the government would undermine its friends and reward any of its parliamentary opposition with an admission that the Mint was badly run. The committee did not condemn Chaloner as a liar and a thief—quite the reverse. But neither they nor anyone else was willing to expend any political capital to force him on a Warden who was clearly determined to keep him out.

The issue came to a head in late spring of 1697, when the parliamentary session ended with no offer of preferment for Chaloner. The news shocked him. Worse, it left him very close to flat broke. His last visit to Newgate would have been as expensive as usual, and he seems to have abstained from any coining enterprises while trying to run his long con on Parliament. But by the end of the winter, "his money grew short," and he acknowledged that "if ye Parliament did not give him encouragement he must go to work again." On March 10 or 11 he commissioned "a stamp for a shilling" from an engraver with whom he had worked before. If he could not trick the government into making him wealthy, he would make his fortune in the familiar way, with what Isaac Newton himself called "a new and dangerous way of coining."

Chaloner now reassembled his old firm. He recruited his longtime co-conspirator Thomas Holloway, and the two of them resumed their partnership, whereby Chaloner supplied the brains and the ambition and Holloway took charge of the logistics.

Desperate for a quick infusion of cash, Chaloner told Holloway to "take a house in the Country, convenient for coyning," while "he should find materials."

Holloway worked quickly, finding a house in the village of Egham, in Surrey, about twenty miles southwest of London. Operations on a scale to satisfy Chaloner needed plenty of space, as they produced a lot of noise and heat along with a constant flux of raw materials, finished goods, and people. Such hubbub could never go unnoticed in London. Every coining scheme in the capital relied on the willed blindness—bought, coerced, or born of indifference—of dozens of witnesses. This *omertà* never held indefinitely. Newton filled his case files—and Newgate—with reports of coining operations in tenement rooms or close-packed houses observed by neighbors or captured small fry who had seen coining apparatus as they came and went with their handfuls of dud crowns or guineas. A rich man's house, either in town or in the country, would not do either. High walls and enough room could defeat the curiosity of strangers, but it was impossible to keep any substantial coining operation secret from the servants any wealthy household would employ.

The choice of a village house evaded both traps. It was private enough to avoid too much local scrutiny. It was modest enough so that no servants need apply; the new tenants would take care of themselves. Best of all, Chaloner and Holloway appear to have believed, it was far enough from London to escape the Warden's immediate notice.

While Holloway worked out the details of the new location, the ringleader handled his chores. Chaloner's "new way" of coining was essentially a variation on the traditional method of casting counterfeits. But his understanding of the demands of high-quality casting seems to have impressed even Isaac Newton, who documented each step of his nemesis's process as he learned it from a parade of informers. The key to casting successful counterfeits lay with the quality of the stamps or molds that impressed the image of the two faces of a coin. To make sure that his molds would pass muster, Chaloner cut the face and reverse patterns into wood blocks and then handed them off to Holloway, who took the patterns to a metalworker named Hicks. Newton caught the essential detail in the next step. Ordinary molds opened to receive molten metal, and reclosing them to produce both the top and bottom faces of the coin could leave suspicious marks. So Chaloner directed Hicks to produce a brass mold that had a channel, or a kind of spout, through which metal could be introduced into the casting chamber—thus, in theory, reducing the likelihood of introducing flaws or telltales into the finished counterfeit.

From Hicks, the brasses now traveled on to a third man, John Peers, who was to file their faces. This step would refine the quality of the image they would leave, making them practically indistinguishable from the faces of coins struck by the Mint's machines. Last, Chaloner insisted on counterfeiting only shillings, which meant that the new molds would be "but little ones . . . so that they might be hidden anywhere."

Early in his career, Chaloner had held close his knowledge of the counterfeiting process, maximizing his take from each dud coin. Now he was more concerned to distance himself from any actual contact with a false coin. So he agreed to teach the Holloway brothers the secrets of his "new way quick and profitable." John Holloway proved an indifferent student, but Thomas showed his quality once again. Chaloner would arrange for someone to pass their bad shillings into circulation, and with this division of labor, "the three should share the profit."

It was a good plan. It should have worked. But within a few weeks, the whole scheme started to fall apart.

On May 18, John Peers—the man Chaloner had chosen to put the finish on his new coining molds—appeared before a magistrate to answer a charge unrelated to the current scheme. When pressed by his interrogator, however, he spilled his guts, volunteering as much as he could of Chaloner's plans. He testified that one of Chaloner's gang had asked him to make an edging tool of the sort used by counterfeiters. He said that he had seen "Cutters and Tooles Instrumts proper for coyning" at a house occupied by Chaloner's brother-in-law, Joseph Gravener (also Grosvenor). Peers admitted his own guilt for providing some of the "divers Tooles necessary" for Gravener's coining ambitions, and he claimed to have seen Gravener "actually counterfeit a Milld Shilling." He said that Chaloner was pressing his brother-in-law to deliver the equipment needed in Egham by promising that he "would have him in a Proclamation about a Fortnight since at Clark's the Flask Tavern"—a deadly threat, as it meant that Gravener would stand publicly accused of a capital crime. Last, and probably most galling, Peers testified that he had heard Chaloner make his famous boast that he would "fun" Parliament as he had previously defrauded the King and the Treasury.

Unfortunately, Peers's information took its own sweet time to reach the man who most needed to hear it. Newton learned of the confession only by accident, three months after the initial deposition. In early August, he visited the Secretary of State's offices to question another counterfeiter in a case unrelated to Chaloner's. There, finally, someone mentioned what Peers had said. The news shocked Newton into action. He arrested Peers on August 13 and brought him to the Tower for questioning. He recognized the obvious, however: nothing in Peers's account directly implicated Chaloner, and nothing very significant had happened yet. Newton needed more, and he knew what he had to do to get it. He released Peers and gave him five shillings for walking-around money, in exchange for which Peers was to report on the doings of Chaloner's gang.

Peers soon ran into trouble. Newton's criminal opposition seem to have noticed his habit of questioning suspects in the Mint, and information about who entered and left by the Tower's western gate became a valued commodity. Within a day, the wrong people knew that Peers had spoken to the Warden. Someone—it is not clear who—denounced him as a counterfeiter to a thief-taker, who promptly delivered Peers to Newgate. Newton had his own sources on the street, however, and word of the arrest reached him almost immediately. He bailed his man out the next day, paying the bill out of his own pocket.

With Peers back on board, Newton proceeded as usual, working his way up the gang to weave the strongest possible net of evidence around his primary target. He was in luck: Thomas Holloway was already in custody, confined to the King's Bench Prison since April for an unpaid debt. Peers visited him, telling him that Gravener had taught him, too, how to use Chaloner's new casting method. Holloway, unsuspecting, sent Peers on to the gang working at the Egham house, and Peers produced eighteen counterfeit shillings, thus proving his willingness to take the risks involved. Chaloner was furious when he heard of the newcomer, blasting the incautious Gravener as "a Rogue for teaching him." But the damage was done. Newton again arrested Holloway, now for coining, and with the danger of the death penalty hanging over him, Chaloner's closest confidant had every reason to talk.

He didn't, at least not at first. But then Newton caught a break. Chaloner grew tired of waiting for the Egham plan to generate a return, so he and another man, Aubrey Price, came up with a new scheme. On August 31, the two men came of their own free will before the Lords Justices to present evidence of what they claimed was a Jacobite conspiracy to attack Dover Castle. They offered to infiltrate the plot as couriers and thus to intercept whatever passed among the notional conspirators.

It was a harebrained notion by any stretch—neither the phantasmagoric quality of the supposed plot nor the credentials of Chaloner and Price as Jacobite thief-takers inspired any confidence. Chaloner must have been either truly desperate for cash or else just overweeningly confident—or he may simply have thought that the worst that could happen was that the justices would say no.

And that's probably what would have happened but for an astoundingly bad bit of luck. On the same day Chaloner tried to sell his story, Newton was giving advice on whether to execute a coiner convicted in an unrelated case. The two men seem almost to have tripped over each other in the halls. Newton recognized Chaloner and identified him to the Lords Justices. The order came back immediately: he was to arrest William Chaloner and prepare the case that would put a final stop to his career. And so, on September 4, 1697, agents of the Warden of the Royal Mint committed Chaloner and Price to Newgate Prison.

Newton had followed his instructions, but he was not a happy man. He knew that the evidence he had gathered against Chaloner so far was perilously thin. In fact, he told the Lords Justices that he did not have enough believable testimony to hold Chaloner for anything more than a misdemeanor. No matter, he was told: let the jury wallow in the gory details of Chaloner's minor offenses, however unrelated to the actual charge. With their minds thus prepared, the justices assured him, London jurors would be ready to return felony convictions on less than airtight evidence. So Newton did as he was commanded, and began to prepare his case for trial.

In the meantime, Chaloner readied his counterattack. At first he merely muddied the waters: he accused Price of being the mastermind of the various plots that might form the basis for a charge. Price returned the favor—and two minor members of the conspiracy also testified, enough to threaten the same kind of muddle that had so damaged the investigation of the theft of the Mint dies. But Chaloner did not place all his trust in mere confusion. He next launched a direct attack on the core of Newton's case.

By far the most dangerous potential witness against him would be the man closest to him, Thomas Holloway. At the time Chaloner was jailed, Holloway had been released, probably in exchange for promised testimony at the upcoming trial. But Newgate was a sieve, and a clever man could reach through its walls and touch those outside. Chaloner turned to Michael Gillingham, the keeper of an alehouse near Charing Cross, who had previously run the kinds of delicate errands that came up in Chaloner's line of work.

On or about the seventh of October, Gillingham met Holloway in his tavern and made him an offer. Chaloner would pay his old friend handsomely—twenty pounds, enough to cover his expenses for several months—if he would only have the good sense to leave for Scotland, beyond the reach of English law. Holloway did not accept immediately. Gillingham kept up the pressure, alternating the carrot and the unstated, but clearly understood, stick: in the past, Chaloner had betrayed men who could endanger him, sending at least two to the gallows. To help Holloway assess his options, Gillingham played benefactor, renting lodgings for his family and promising to take care of his children for the five or six weeks before they could be sent to join their parents in Scotland. When Holloway demanded assurances, Gillingham brought in as guarantor Henry Saunders, a tallow dealer both men knew and apparently trusted.

Finally, Holloway agreed to make a run for it. Gillingham gave him no time to think twice. He handed Holloway nine pounds on the spot. He paid another three pounds to Skipper Lawes, master of the ship that would carry Holloway's children to Scotland. A few days later, Saunders again accompanied Gillingham as Chaloner's agent. Holloway handed Gillingham a document empowering him to collect some debts owed him, and then he and his wife mounted two horses hired for the journey north.

There was one last detail to square away: Holloway had told the livery owner that he intended to return the rented horses that night—but Gillingham knew that this was just one more casual fraud. He went to the livery stable in Coleman Street and "told the man of the house that his horses would not be at home till two or three dayes were over"—news that must have cost him the extra hire. Why did he bother? Because he "would not have Holloway persued by the man for his horses."

Then, with every thread meticulously woven up, Gillingham visited Newgate to report to his client—still with the useful Harry Saunders in tow. Chaloner "asked him if Holloway was gone away." Gillingham said he was—which had the desired effect, as Saunders reported: "Chaloner then seemed to be Very joyfull and said a fart for ye world."

Newton's premonition of fiasco was justified. With Holloway nowhere to be found, the two other witnesses recanted, although what the defendant did to produce such sudden amnesia is unknown. The case never reached the jury; the presiding judge dismissed the charges. By the end of October or the beginning of November, after seven weeks in jail—in irons, he claimed—Chaloner walked out of Newgate, a free man once more.

Part VI

---◦◦◦◦---

Newton and the Counterfeiter

19

"To Accuse and Vilify the Mint"

FREE CHALONER MIGHT have been, but he was a deeply worried man. By December 1697, he was virtually destitute. Keeping up the appearance of respectability before Parliament had left him short of cash. Pile on a seven-week stay in Newgate, and the cupboard was bare.

With winter approaching, trying to live on what his jailers had left him goaded Chaloner to the point of recklessness. Had not an English judge refused even to present the trumped-up case against him to an English jury? Had he not suffered the shackles, the squalor, the naked corruption of Newgate? Should not someone compensate this guiltless man for all the wrongs done to him?

On February 19, 1698, Chaloner laid his portrait of abused virtue before Parliament in a document he also had printed for public distribution. "Your Petitioner," he wrote, "did in the last sessions of Parliamt discover several abuses committed in the Mint." And what was his reward for such service to the Crown? "Some of the Mint threatned by some means to prosecute him & take away his life before the next sessions of Parliament." His accusers had gone so far as to conspire with the worst kind of scum to suborn the crime that would bring him to his death: "some of the Mint have imployed & given Privilege to several persons to coyn false money . . . all of which was done with an intent to draw him [Chaloner] into coyning to take his life away."

This attempt at judicial murder failed, defeated by Chaloner's determined virtue: he was concerned only "to find out the Treasons & Conspiracies against the King & Kingdome" and then "this year writing a book of the present state of the Mint & the defects thereof . . . wch he hoped would have been of service to the Publick." That the Mint would not abide, of course, and so, Chaloner charged, "they committed him to Prison & so prevented him from doing it."

The miseries of the cells had brought him "great sufferings & ruined condition," and left him "incapable of providing himself & family." There must be someone to make him whole, or as Chaloner humbly put it, should give him "such redress as shall seem best in your Honours great Wisdom & Justice."

There could be no doubt whom Chaloner really meant by that careful phrase "some of the Mint." Isaac Newton was the only man who had both means and motive to use the power of the state to kill a man for private revenge. Newton himself certainly understood. He copied out Chaloner's petition in his own hand, and four versions of his reply survive in his papers. Bitter anger runs through all of them, along with a healthy dose of disdain: "If he would be let the money & Government alone & return to his trade of Jappaning," Newton wrote in his first attempt at an answer, "he is not so far ruined but that he may still live as well as he did seven years ago when he left of that trade & raised himself by coyning."

Yet an odd, pleading tone also pervades each of the drafts. The problem was that Chaloner was telling the truth, more or less. Witnesses had failed to appear. No link had been shown between the coining den in Egham and Chaloner himself. The case—as Newton had feared—was laughably weak. His complaint that Chaloner had "laboured to accuse and vilify the Mint" looked like confirmation that Chaloner's arrest was ordered out of injured pride. His declaration that there were "divers witnesses that Mr Chaloner last spring & Summer was forward to Coyn" was true but beside the point, given that none of those witnesses proved willing to show up in open court. And when he complained without proof that the defendant was guilty of the kind of witness tampering that "gravells prosecutions & renders it dangerous for any man to prosecute," he simply sounded weak in the face of an opponent who had bested him.

It got worse. Newton added: "I do not know or beleive that any privilege or direction was given by any of the Mint to draw him or his confederates in." That phrasing sounds just a bit too careful a dodge—and it was, for of course it was Newton himself who had given John Peers money and sent him off to infiltrate the Egham gang—even bailing Peers out of Newgate to do so. Here he seemed to be looking for plausible deniability if Peers or any of his other agents should turn up to confirm Chaloner's tale.

Chaloner's petition sparked yet another official investigation, and for the moment roles were reversed: Isaac Newton was standing in the dock, defending himself against the charge of framing an innocent man. A panel of senior government figures was assembled to look into the matter, and though the group was stacked with Newton's friends—Charles Montague and such reliable allies as Lowndes and James Vernon, then serving as Secretary of State—initially the evidence heard by the group, including Chaloner's own testimony, tended to favor Chaloner's claim. The panel persisted, however, and as other witnesses testified, more and more gaps turned up in the plaintiff's story. In the end, the investigators produced a report that dismissed Chaloner's claims—but quickly, in a bald rejection that did not satisfy Newton's hunger for a full exoneration.

But if Newton felt aggrieved at this perceived slight—the more so, perhaps, for having been so nearly caught out—he knew who had truly caused him such vexation. He was certain that Chaloner had committed crimes against the King, and that was bad enough. And now he had formed "a confederacy against the Warden."

This was new: Chaloner had been just one more anonymous offender, against whom equally interchangeable officials would take the steps needed to cut short a criminal career. But no longer. This one criminal had targeted a single, specific officer, the Warden. Alone of all those he sent to Newgate and the gallows in his years as the coiners' scourge, the Warden of the Mint did William Chaloner the honor of treating him as an individual antagonist—someone not merely to be stopped, but crushed.

The ruthlessness to come in the pursuit of Chaloner had deeper roots than mere anger over the humiliation of having to defend himself in public. Newton had already proved willing to pursue ends over means when he acquiesced in the Lords Justices' suggestion to so prejudice the jury as to extract a felony conviction for misdemeanor offenses. But the ferocity he showed through the next phase of his campaign against Chaloner suggests that there may have been more than mere *raisons d'état* driving him. Chaloner could not have known that there was a hidden thrust concealed within his challenge to the Warden, one that touched Newton's most private faith.

Faith indeed, for any counterfeit had religious significance. The magic that transformed a disc of metal into legal tender came from the image of the King's head on the face of a coin. The King ruled by the grace of God. To steal that likeness was an act of lèse majesté, an offense against the sacred person of the monarch. Coining was a capital crime because of the danger it presented to the state; it ascended to the odium of treason because of its insult to the Crown.

But while that was true for any counterfeiter, Chaloner had mocked not just King William III but also Isaac Newton, and on very specific ground. By 1698, Newton was no longer a practicing alchemist. Still, Chaloner's counterfeiting was, in effect, a blasphemous parody of the alchemist's dream to multiply gold without limit—the equivalent of a black mass, in which a toad or turnip takes the place of the consecrated Host. The same would have held true for any forger, of course. Yet none but Chaloner ever set himself up as a direct rival to Newton's mastery over metal.

Did that trespass matter? Did Newton pursue Chaloner more intensely than he would have absent his own alchemical history? It is impossible to know. Clearly, Newton's motives for hounding his quarry were overdetermined: duty and personal offense as well as any secret defense of faith all fed into the mix.

It is important to remember, however, that while many of his biographers have drawn portraits of a swarm of different Newtons—the magician, the mathematician, the experimental genius, the young Newton as a cloistered professor, the older man in charge of the Royal Society, conducting the running war with intellectual enemies on the Continent—the real Isaac Newton was one man living one life, whose parts as he lived them were thoroughly conformable to the whole. Within each role, every job he did, each problem he set himself, that one Newton remained—and the constant theme of that singular life was his hunger for contact with the Godhead.

That same man understood the disquieting fact that the new science, in the wrong hands, had the potential not to prepare men for "beliefe in the Deity," but to undermine their faith. Into that knowledge enter Chaloner, whose every action reeks of a kind of practical atheism: what need for God to act in the world when a smooth enough operator can produce passable imitations of His works?

Whatever its precise root, the fact is that following Chaloner's release in February, Newton's anger was never more intense. From that moment, the Warden of the Mint pursued single-mindedly and relentlessly the man who had managed to offend him in every conceivable way.

20

∾⟡∾

"At This Rate the Nation
May Be Imposed Upon"

FOR ALL OF Newton's secret rage, William Chaloner had more pressing problems—or he thought he did—in the spring of 1698. He remained very poor—perhaps the one wholly true statement in his last petition to Parliament. It had been more than a year since any of his schemes had produced a real return. The gang he had assembled the previous summer had been broken up, its members jailed or fled. His parliamentary allies had failed him; whatever else had happened, he recognized that the faction supporting the existing administration of the Mint—Newton—had prevailed over those who had backed him as part of their campaign to return to power.

Compounding his difficulties, even counterfeiters need money to make money, and Chaloner had no capital left. In June he set out to lever himself out of his predicament, starting by making a few crude shillings—enough, perhaps, to fund more ambitious plans. The whole mean affair revealed just how far Chaloner had tumbled down the criminal ladder. The Knightsbridge house was long gone, and he now lived in a rented room over the Golden Lyon on Great Wild Street, near Covent Garden. No longer able to keep physical distance between himself and his coining operation, he did what he could over the fire in his own grate. A witness reported that he watched Chaloner "bring out of his Clossett 2 pieces of white earth like Tobacco pipe Clay of the consistence of Dough or paste." Chaloner had stuck a shilling coin between the two pieces of raw clay, and "pulling them asunder he took out the Shilling and laid the pieces of earth to dry by a fire and ... at last upon the fire to dry and bake them throughly and after they were could [cold] they rung like burnt earthen ware."

This was almost literally child's play, the kind of slipshod work that routinely doomed amateurs making their first attempts. Chaloner knew better, of course, but he simply could not afford the fine craftsmanship that he had previously used to keep himself out of trouble. Instead, he persuaded his old partner Thomas Carter to give him three shillings, and then "meltd them down with some pewter and Spelter in an Iron Ladle and cast abot. ½ a Score Shillings in the earthen mould." It was a tedious exercise: "he cast but one at a time and as often as he cast a bad one he threw it back into the Iron Ladle." This was no way to get rich.

Even at this low ebb, Chaloner did not try to pass his crude fakes himself. But Carter would not take them either, "being afraid to have them abot. him." The next day, Chaloner tried them out on a metal dealer, John Abbot, who had not been above selling Chaloner silver and gold in better days, but the new fakes did not pass muster with him either, because "they were so light." A week went by, and Carter finally agreed to see what he could do. He sent his maid, Mary Ball, to pick up half a dozen samples. Chaloner handed them over, telling Ball that if Carter "did not like them he would do some better." Then, as quickly as he could, he turned to a new and much more profitable opportunity.

This time it was King William's seemingly endless war that gave him his opening. The cost of the war, combined with the shortfall in tax takings created by the coinage mess, had forced the government to experiment with almost any gimmick that anyone could invent to raise funds. With sixty thousand or so soldiers tramping through Flanders, neither the Bank of

England's notes and checks nor the early form of government debt called Exchequer bills raised sufficient funds. To fill the gap, clever men cast about for ever more exotic financial ideas. Of them all, perhaps none was as strange as the Malt Lottery scheme.

The Malt Lottery was actually the sequel to an earlier attempt to play on the English love of a flutter, a scheme put together three years earlier by someone who understood the get-rich-quick impulse all too well: the Master of the Mint, Thomas Neale. In 1694, Neale's "Million Adventure" had offered 100,000 tickets priced at ten pounds, each with a twenty-to-one shot at prizes that ranged from ten to one thousand pounds. Even better, each ticket carried with it a sweet interest payment: one pound a year through 1710, for a guaranteed minimum return on the "Adventure" of sixteen pounds.

Neale did well for himself on the deal. He kept ten percent of the proceeds, a fairly modest cut by the standards of the day. (Promoters of a similar lottery in Venice skimmed one-third of the receipts off the top.) He knew his customers too. He kept the price of tickets low to attract "many Thousands who only have small sums and cannot now bring them into the Publick, to engage themselves in this Fund." In fact, while at ten pounds apiece whole tickets were still too pricey for most, they were just cheap enough for speculators to buy and syndicate, selling shares in tickets to the kind of democratized financial customer Neale had in mind.

It all worked, for a while. The Adventure sold out to a much wider range of people than had ever before voluntarily lent their money to the state—tens of thousands of individual investors, according to some estimates. The diarist Narcissus Luttrell recorded that a stonecutter named Mr. Gibbs and his three partners took home one of the five-hundred-pound prizes, and a Mr. Proctor and a Mr. Skinner, a stationer and a hosier, respectively, won another. These were hardly the usual sort of people involved in high finance, but here they were, new men and women seeking, and occasionally finding, wealth in these strange new forms of money. The Adventure pushed further into unknown territory after the prizes were distributed shortly after the initial sale of the tickets, with the emergence of an informal bond market. Traders who had waited until the lottery prizes had been distributed bought tickets below par—paying as little as seven pounds, or seventy percent of face value—to secure the best possible return on their capital.

It all ended in tears, though. The government had got its sums wrong. The Adventure tickets carried an impressive rate of return, over and above the prizes paid out to the lottery winners. Unsurprisingly, the cash-poor Treasury had trouble meeting its payments. The first signs of a shortfall came as early as 1695, less than a year into the putative sixteen-year life of the ticket-bonds, and by 1697, the fund that was supposed to pay off the notes was running almost a quarter of a million pounds behind its obligations. Angry ticket holders petitioned Parliament, demanding that the government defend the "Credit and Honour of the Nation," but there simply was not enough money on hand to resume payments until peace finally came in 1698.

Before that, as the war ground on, King William's army needed yet more money. It made sense—at least to Neale—to try another lottery. To deal with any lingering unpleasantness over the default on the Million Adventure, Neale tied this new lottery to the excise tax on malt (in effect, a tax on beer). The Malt Lottery opened for business on April 14, 1697. The Treasury issued 140,000 tickets, with a face value of ten pounds each, to be sold to the public. Those tickets, which were supposed to bring in 1.4 million pounds, were chimeras: part bond, part bet, part paper money. The lottery promised cash prizes and an ongoing stream of interest, just like the earlier venture, but the new tickets were not merely bonds, to be bought and sold like any other investment. They were actual bills of exchange, cash, legal tender from the moment they hit the London streets.

At least that's how it was supposed to work. Neale apparently believed that the thrill of the game, combined with the fact that these new tickets could pass for cash, would overcome the public's lack of faith in their ever more indebted government. The Chancellor of the Exchequer, Newton's old patron Montague, had his doubts, complaining that "nobody does or will understand the Lottery Tickets and the Merchants will not meddle with them." Montague was right. No one trusted the strange new paper, and in the end, just 1,763 tickets found their way into the hands of the public.

Nonetheless, the government desperately needed the money the lottery was meant to raise. So the Treasury simply treated the remaining 138,237 tickets as ten-pound notes—cash held by the government, to be paid out to anyone who could be forced to accept them. Astonishingly, it worked—sort of. In accounts for 1698, the Royal Navy reported that it held almost forty-five thousand pounds' worth of Malt Lottery tickets to cover the pay due to sailors and marines—exactly the kind of captive creditors who did not have a lot of choice in the matter.

So it happened that, without admitting it, the Treasury invented a parallel English currency to the bits of metal still passing from hand to hand. This new paper was not the equivalent of true fiat currency. The fact that the issue was tied to a specific asset, the revenue stream from the tax on malt, gave it a hybrid character as both cash and secured debt. But if it was not quite the same as modern paper notes, it was still radically unlike anything Englishmen had ever known as money.

That was as far as it went for the time being; lotteries did not play a significant role in government finance in the years after the failure of the Malt issue. But it had become obvious not just to the King's bankers, but to almost everyone—stonecutters, maids, hosiers—that the nation's financial system had not kept up with what was actually happening in England's economy. Chaloner grasped much more quickly than most that hard cash—the material reality of silver and gold—was no longer the only, or even the most important, form that money could take.

Among the great mass of men and women outside the learned societies, the scientific revolution was making inroads on understanding as the world of money came fully into being. Paper money, exchangeable promises, bonds, and loans are all abstractions. To understand them, to accept them—even to suborn them—took a capacity for the kind of mathematical reasoning that was just beginning to infiltrate all kinds of new ideas, including that demanded by the new physics. Figuring out the present value of a bond, for example, or how to price the risk (likelihood) of government default, demanded and demands a quantitative, mathematical turn of mind—just as calculating the orbit of a comet did and does. William Chaloner was hardly a scientific revolutionary. But he recognized that a revolution was happening around him, and he had wit enough to seize the opportunities such radical changes in thought and practice created.

Chaloner spent the month of June, 1698, trying to work out how to profit from the Malt Lottery without putting himself in too much danger. To get back into competition with the government he needed some tools: a properly engraved plate, the right ink, and the correct paper—much less than it took to set up a profitable criminal mint. But the blunt, brute fact of his poverty held Chaloner in a vise. Even the modest amount required to set up a print shop was beyond him. He needed help, but he had been as profligate of friends as he had of money. His most trusted partner, Thomas Holloway, was long gone, pushed into Scottish exile. And now there were only a few men left who had worked with him in the fat days of the early 1690s. Ultimately he chose to open his mind to one of them, his occasional collaborator and the reluctant receiver of his bad shillings, Thomas Carter.

Carter told Chaloner that he knew someone with money who was willing to fund counterfeiting schemes. Chaloner was skittish. He knew—none better—how easy it was to make a good living betraying one's co-conspirators. But he had no real alternative, and so he told Carter that he could make contact as long as he kept Chaloner's name out of the proposition.

Near the end of June, Carter met his man, David Davis, promenading on Piccadilly. They talked—apparently out in the open, on the street—for some time. At last Carter stated his business. Davis said Carter told him "he was acquainted with a man which wuld engrave very dexterously," and that his friend had "a Strong inclination to grave a Plate for Malt Tickets." All he needed was a bit of backing, Carter promised, and he and his friends would be rich men.

Carter did his best by Chaloner, warning Davis that he "must ask noe questions," but Davis pressed him harder, saying "no person yt I ever heard off did understand taking the reverse of a fine bill upon Copper besides Chaloner." Carter replied that "if you knew who my friend was you would allow him as great a Mastter as Chalon(er)"—which was, after all, no lie. Finally they agreed: Davis would give Carter money and sample Malt tickets to guide his mystery engraver. In return, Carter promised Davis daily updates until the plate was finished.

It took Chaloner several weeks to complete the fine work, engraving an exact copy of each side of a ticket onto a sheet of copper plates purchased with Davis's seed cash. Carter kept his end of the bargain, providing "an accot. of the every day's proceedings until the plate was finisht." He did make one mistake, however: at some point he allowed Davis to discover the identity of the third man in the scheme. As Davis later boasted, "All this time I had sufficient assurances that Chaloner was the person which engraved the plate."

Davis was, inevitably, the Judas that Chaloner had feared. He reported to a man with whom he had done business in the past: not Isaac Newton but the Secretary of State, James Vernon. In his true guise as a paid informer and thief-taker, Davis told Vernon that Chaloner had finished the plate. Warning that time was of the essence, Davis asked Vernon for one hundred pounds — right away, he said, "to prevent the destrubutions fo Severall falce Tickets . . . and to subsist the persons that had done ym." He told Vernon that he would buy up the entire print run "till I could obtain ye advantage of Seizing Chaloner and of Securing the Plate." Mindful of his own interests, he kept Vernon in the dark as much as he could, revealing neither where he met Carter nor where Chaloner lived.

It took Davis two more interviews to persuade Vernon to hand over the hundred pounds. Cash in hand, Davis told Carter that he had found a buyer for two hundred tickets — two thousand pounds face value — and thus needed him "to let me have all the Counterfeits that were taken off the plate." In return, Carter "& his friend should have continuall supplys till they should print all the rest."

Carter took the bait, handing over a packet of fake tickets, with which Davis, "having thus secured all wch. I understood were printed," returned to Secretary Vernon. The work was good enough to terrify the authorities. Vernon ordered Davis to be "very industrious in finding out Chaloner" — and, above all, to locate the engine of the crime, his meticulously crafted copper plate. Luckily, Carter was now so far gone in gratitude to his buyer that he seems to have been unable to stop talking. That was how Davis learned that Chaloner hid the plate in a wall between printing sessions. But which wall?

Davis found himself caught between Vernon's pressure and

Chaloner's sense of self-preservation. He played the only card he could, sticking to Carter like the man's own clothes, except for when Carter had to pass supplies to and receive finished goods from his partner. All the while, Chaloner bent to his task. Carter reported that his partner meant to keep printing until the plate wore out. And so it went: Vernon pressed Davis; Davis harassed Carter; Carter begged Chaloner to let him see the plate; Chaloner refused.

At that pass, all of the weeks of pursuit crashed to a stop. Vernon's surveillance ran into the old problem: the right hand knew not what the left was doing. While Davis followed his line, another, completely separate operation was being run out of the Tower. Isaac Newton had not forgotten William Chaloner, not for a moment. Davis's exclusive grip on one informant, Carter, had kept news of the Malt scheme from reaching the Tower. But Newton was still working through the ruin of the Egham coining case. His prosecution had collapsed when Thomas Holloway ran to Scotland. But even if the Warden could not compel obedience across the border, he could persuade—and in early autumn 1698, Chaloner learned some deeply unwelcome news. Newton had found Holloway, and his old partner was prepared to cooperate with the Warden. Chaloner reacted immediately to the danger. He dove for the weeds, shutting down the Malt ticket production line until he could gauge the danger posed by Holloway's return.

Davis knew what had happened, thanks to Carter. But he kept Vernon in the dark. It was hardly in his interest to hand over any information that could make him less valuable to his patron. Consequently, neither Vernon nor Newton knew that their investigations had just collided. Vernon's case was running ahead of Newton's, thanks to Davis's ability to play Carter for a fool—but the plate was still at large, and the man who knew how to use it could not be found.

Worse, Vernon's office was leaking secrets. Chaloner learned from a man named Edwards that Davis had been seen in the Secretary of State's presence, promising to deliver up a counterfeit plate. Carter at last grew suspicious of his too helpful buyer, while Chaloner made sure that all physical evidence of his crime disappeared. With that, the case teetered on the verge of embarrassing failure.

Davis did what he could to calm his skittish contact, reminding Carter that Edwards had cheated him out of some money. As that memory festered, Carter, incredibly, began to confide in Davis again. But the weeks were passing—it was late October—and Davis had to admit to Vernon that he still had no idea where Chaloner had hidden his tools. Vernon took the news badly, "full of displeasure saying at this rate the nation may be imposed upon."

Davis understood. He promised Vernon that he would catch Chaloner in the act within a week, or else abandon the effort (and any profit) to "his hon[our's] discretion." Chaloner was still ahead of the pursuit, though. Carter reported that the plate was now hidden with a local midwife. She in turn took it with her thirty miles from London, beyond Davis's reach.

Four days passed in the promised week. Davis hammered Carter for news, but all he got was a story that changed by the hour. Chaloner had said he would retrieve the plate himself, and soon. Then no—Chaloner promised to send a messenger for the plate, which would arrive in London the next morning. The printing, presumably, would start up as soon as the plate was restored.

Suddenly it all fell apart. Carter had deceived Davis when he told him that he had handed over all the tickets Chaloner could print—or, more likely, Chaloner had cheated the ever-gullible Carter, selling some tickets out a side door. In either case, that same week a man named Catchmead pawned a packet of false lottery tickets—Chaloner's finest work—for ten pounds. The pawnbroker in turn tried to pass some of the fakes, and was arrested on the same day Carter told Davis of the impending return of the plate. "The news did astonish me," Davis wrote—which was probably true, given how much his ignorance cost him. To cover his failure, he rushed to the Secretary of State's office.

Vernon was out. Several hours passed before the two met. When they did, Vernon raced to regain control of the fiasco. Davis could no longer keep his secrets. It was intolerable to imagine that London might be flooded with thousands of pounds' worth of false Malt Lottery tickets. Vernon offered Davis this one kindness: he was permitted to capture the smaller and less valuable prize, Thomas Carter. The mastermind, though, was now anyone's game. The Secretary of State's office put a healthy price on Chaloner's head: fifty pounds—a useful sum, enough to keep a family in tolerable comfort for a year—to the man who brought him in, with or without the plate, as soon as possible.

London, that enormous city, could become a terribly small town. Chaloner had made himself as invisible as possible throughout the spring and summer. But he still had to eat, buy beer, find a room. He was known, at least a little, by at least a few. It was enough. Once his person became valuable enough to make it worthwhile to find him, it was just a question of how quickly he would be collared by one or another of London's thief-takers. The size of the reward guaranteed it would not take long. Davis lost out. Frustratingly, the record of Chaloner's arrival in custody does not reveal how he was found or where he was taken. All that can be known for sure is that within days after Vernon opened the season on Chaloner, he was carried to Newgate by a man named Robert Morris, who had in the past tracked men for the Mint.

Once Vernon had his man in jail, Newton finally learned of the parallel investigation. Though the Warden of the Mint had no official reason to worry about the Malt Lottery—that was the Treasury's problem—he managed to persuade Vernon to let him take the case. With that, every intermediary dropped away. The game was down to its essence: Isaac Newton versus William Chaloner.

21

⸺◦◇◦⸺

"He Had Got His Business Done"

IN THIS SECOND ROUND, Isaac Newton took no chances. He made sure that the jailers held his man close. Through November and December 1698, Chaloner found himself as isolated as his captors could make him. From his cell, the accused complained that the only visitor he had been permitted was his small child, adding piteously, "why am I so strictly confined I do not know."

Jail did not rob Chaloner of his confidence, though. The Malt ticket plate remained hidden, and Chaloner claimed he had nothing to do with it anyway. When he was taken, he had no false tickets on his person. There was Carter's testimony to consider. But faithful, garrulous Thomas Carter—now also lodged at Newgate—was the only member of the alleged conspiracy known to have been paid for passing the counterfeits. If anyone were to fall for that crime, Chaloner was sure it would not be he: in his cell "he made very slight of the matter, bragging that he had a Trick left yet."

It pleased Newton to let him think so. The Warden had learned from the debacle of the previous year. Even before Vernon's reward brought Chaloner in, Newton had begun to reconstruct his adversary's entire career. Most of what he had learned was background, not really actionable, but useful nonetheless. For example, in May 1698, Edward Ivy (aka Ivey, aka Ivie, aka Jones) swore before Newton to having direct knowledge of an impressive range of currency criminals. Ivy was a confidant of John Jennings, one of the Earl of Monmouth's footmen, who traded in high-quality false currency; he knew Edward Brady, who "made it his constant business to utter counterfeit Gineas"; he testified against Whitehall's famous porter, the vicious John Gibbons—who, Ivy claimed, had conspired with Brady in the occasional highway robbery. He knew John and Mary Hicks and their daughter, Mary Huett, who together ran a family business clipping the old currency; he was willing to reveal name after name: "one Jacob," Samuel Jackson, George Emerson, Joseph Horster, "and other Emint. Coynes and Clippers."

William Chaloner played only a minor role in Ivy's catalogue. Ivy mentioned the object of Newton's interest just once, when he testified that he had asked Jennings if his fakes were as good as Chaloner's, and Jennings said they were, that "Chaloner was a fool to him that made the said Gineas." Ivy added that he believed —but clearly was unwilling to swear that he knew—that Brady had received some of his supply of false guineas from Chaloner.

Newton took dozens of depositions like Ivy's, at first concentrating on quantity rather than quality. Most of the testimony he gathered through the spring and summer of 1698 was hearsay. Many of the depositions devolved into lists of all the coiners and crimes the witness could remember. Some implicated Chaloner, some did not, but Newton was accumulating a picture of London's coining ecosystem. He was gathering names and noting the links, the web of criminal connections within which Chaloner himself had to move.

Over the months, more and more of the people to whom those names attached turned up in their own depositions—which meant, in practice, they waited on Newton's pleasure in Newgate or some other jail. Those scattered references led to more witnesses, and then to yet more—a scaffolding of information received on which Newton planned to hang William Chaloner.

By January 1699 Newton was spending almost all of his working days at the Mint, conducting the interrogations that would form the heart of the prosecution. By February, his commitment had become total—at one point he spent ten days in a row questioning witnesses. The record is far from complete, but more than 140 surviving statements give a sense of his procedure.

The form of Newton's interrogations was always much the same. Most began by identifying the witness, usually by trade or profession and parish, though some of the women were identified merely as wives or companions of other targets of Newton's inquiries. Newton's questions do not survive, but his approach seems to have been more or less chronological: when did the deponent meet Chaloner; what crimes did he or she witness or hear of, and in what order. The witnesses all talked, often at length, telling tales of crimes the better part of a decade gone by—every detail they could remember, and perhaps some invented to satisfy the terrifyingly persistent man who bent to every word. When Newton was done, he would dictate a summary of what he had heard to a clerk. Either Newton or the clerk would read back this gloss on the testimony to each witness, who could then add to or alter the account. Once both were satisfied, Newton and his witness signed the document and the clerk would produce a fair copy to be entered into the Mint's records.

Over time, Newton found that some of his best leads came from the wives or mistresses of the men Chaloner had partnered with and betrayed. Elizabeth Ivy, identified only as "Widdow"—presumably Edward's—said she had known Chaloner to make false coins at the very start of his career. More important, so did Katherine Coffee, wife of Patrick Coffee, the goldsmith who had first taught Chaloner the rudiments of coining.

Here was the kind of testimony a jury loved to hear: that of an eyewitness who had observed an actual criminal act. Katherine Coffee swore that "abot. 7 or 8 years ago she hath seen Will[iam] Chaloner now prison[er] in Newgate often coyn French pistolls with stamps and a hammer in Oat Lane by

Noble street up 3 pair of staires." Katherine Matthews, Thomas Carter's wife, backed up her husband's stories with a meticulous memory for detail. She had seen with her own eyes, she said, Chaloner gilding false guineas "at the lodging she had hired for him at Mr. Clarkes behind Westmr. Abby." What's more, she had held those coins in her hand, "abot 10 of thes counterfeit Gineas from Chaloner, and gave him 8s a piece for them."

The parade of witnesses lengthened, and with it the catalogue of incriminating testimony. A Humphrey Hanwell added details to the story of the pistoles; Chaloner hammered them out of silver, he said, to produce coins that could be gilded by both Coffee and "one Hitchcock." Hanwell went further, adding that he had seen Chaloner clip coins in the late 1680s, and that soon after, Chaloner had showed him counterpunches for making shillings and "either Ginea Dyes or half Crown Dyes but which the Depont doth not now rem[em]ber."

This last may have been a fantasy, or rather, a desperate attempt to please the interrogator. If pressed, Hanwell would probably have connected Chaloner to Monmouth's uprising, the Gunpowder Plot, and perhaps even to the archer who pierced Saxon King Harald's eye. For his part, Newton was by now experienced enough not to believe everything he was told. In his summary dossier of the investigation, a document he titled "Chaloner Case," he emphasized the Coffee-Chaloner connection and the manufacture of the pistoles as the first coining crime to be laid at his prisoner's feet, passing over in silence Hanwell's wilder claims.

Newton persisted, moving on to the central witness in Chaloner's most recent crimes. In January, Thomas Carter told Newton that while he was at work on the Malt Lottery scheme, the metal trader John Abbot had conspired with Chaloner to make better versions of the pewter shillings that had failed to pass muster the previous June. Two weeks later, Newton hauled Abbot into the Tower, and Abbot poured out everything he knew:

Chaloner had shown him his set of coining dies; Chaloner had bought silver from him as raw stock for false guineas; Chaloner had once told him that he and his brother-in-law produced six hundred pounds of false half-crowns in just nine weeks.

And so it went: Elizabeth Holloway finally revealed the whole intricate story of her family's journey to Scotland, which had allowed Chaloner to escape Newton's first prosecution. Consistent to the end, she revealed, Chaloner had cheated her husband, paying him a dozen pounds instead of the promised twenty. (According to Elizabeth, the sea captain contracted to carry the Holloway children north also got shorted eleven shillings out of a fare of three pounds eleven shillings.)

Newton pushed on, voracious, almost indiscriminate. Cecilia Labree, in Newgate awaiting execution, was urged by a friend to "save her self by confession"—admission, that is, of more than her own crimes. So "for making her confession more effectuall," her friend told her that Chaloner and a confederate "had then a Coyning Press at Chiswick" and that they "were then concerned togeather in making Gineas there." Labree followed that advice, trotting out the story for Newton. The gambit did not save her—she was condemned to die later in 1699—but it added to the pile of similar accounts Newton was compiling.

His approach was taking shape: the particulars provided by any given witness mattered less than establishing that an army of men and women were prepared to say that they had seen/helped/heard of Chaloner making shillings/half-crowns/crowns/pistoles/guineas seven years ago, or five, or three years past, or last summer. Newton was making sure he could overwhelm any jury, to the point where the details of exactly what happened when simply wouldn't matter.

22

"If S^r Be Pleased . . ."

OF COURSE, ALL THE while Newton was building his case, William Chaloner could hardly be expected to remain idle in his own defense. He knew that Newton was pressuring his fellow inmate Thomas Carter, who in turn was well aware that he was almost certainly doomed. Carter had, after all, been caught red-handed, with witnesses to place the forged Malt tickets directly in his hands. Clearly, his only hope was to trade his life for someone whose skin was more valuable than his own. Both Chaloner and Carter were housed on the Masters side of the prison. It was impossible to keep the two men from speaking from time to time, and Chaloner took full advantage of the opportunity to push Carter to act against his self-interest.

He started by trying to persuade Carter "to joyn with him representing if they joined they should save themselves," pressing Carter so relentlessly that Carter asked one of his jailers "to forbid his access to me." Chaloner's message was unwavering: what was past was past, but now they were a team, or as he told Carter, "Wee have played the fool one with another hitherto for want of an understanding betwixt us but now if you'l joyn with me nothing can hurt us & weel fun them all."

That was the carrot. Chaloner also wielded a stick. He wrote to Secretary Vernon, betraying Carter as a repeat offender, formerly confined "in most Gaols in England" for coining, housebreaking, and forgery—not to mention his seven times in the pillory. Carter was beside himself when he found out that Chaloner had attempted to undercut his value as a witness. "I askt him," Carter wrote, "why he informed yo[ur] hono[ur] that I was outlawed." The answer was obvious: "he said because I should not be an evidence agt. him." With that, Chaloner seems to have believed that he'd neutralized his former friend, thus gutting the charges based on the Malt ticket scheme.

He misjudged his man. Newton had been there before him. From the moment he returned to Newgate, Chaloner never lacked company, fellow prisoners sharing his quarters. At least three of his cellmates were working for the Warden, and first among them was Thomas Carter.

Carter had done his best to get into Newton's good graces, adding to his Malt Lottery confession the details of half a dozen coining schemes he had joined over the past several years. But Newton wanted more—testimony against Chaloner of hitherto unsuspected crimes, in terms more persuasive than the testimony of one member of a gang against another—and that Carter could not give him without help.

Enter John Whitfield, last overheard murderously and treasonously promising death to Isaac Newton "if ever King James came again." Then, it had been the debtor surgeon Samuel Bond who retailed the remark to establish his credit with the Warden. Now it was Whitfield's turn to try to eavesdrop his way to Newton's mercy. At Newton's command, Carter ran Whitfield. His instructions were simple: get close enough to Chaloner to hear whatever he might say—especially the location of the missing Malt ticket plate. In early February, Whitfield wrote to Newton, "I have managed that affaire and I think it may be to the satisfaction of all that hear it." But he declined to commit whatever he knew to paper, angling for a meeting with his handler instead. Let him be conveyed out of Newgate, and then, if "S^r be pleased to come to the Dogg . . . I do not doubt but I may be worth yor while and some find hiden Treasures lye in Concaves where never found by reason."

And so it happened that Isaac Newton, Warden of His Majesty's Mint, took himself to the Dogg pub and had Whitfield brought to him there. Whitfield told him that Chaloner had hidden the plate in some hollow in one of the buildings to which he had access in the last week or so of the conspiracy. But he did not know where the Mint's men should search, nor even in which house they should start. He could tell Newton only that "it was never lookt for in such vacan[t] places."

Newton did not leave a record of the meeting, so there is no way to tell how annoyed he was by this bait and switch. The sequence of events suggests that he remained patient. Whitfield returned to Newgate, apparently with instructions to continue insinuating himself into Chaloner's confidence. He failed. Chaloner still had some contacts at large in London, and Thomas Carter reported, "After Mr Whitfield was with you at the Dogg, Chaloner was a little suspicious of him."

A game of cat and mouse followed. Chaloner, famous for the love of his own voice, now had to learn how to shut up. To an extent, he did. What Carter had tried to paint as a temporary setback for Whitfield was a complete rejection. "Mr. Whitfield has endeavoured all he can to get more out of him as farr as is conven[ien]t," Carter confessed to Newton, but Chaloner would not be drawn. "All his discourse to him is that he hopes he is a man of honour and will not talk of any thing els[e] to him." That was bad news for Whitfield—Newton valued results much more than effort—and hence bad for Carter too. Carter pleaded with his captor for one more chance. With Whitfield useless, Carter told Newton, "I hope I have got as good an Instrumt. If you please to think well of it I will get out of him all that he has done and all that he intends."

Newton did. Carter's new man was John Ignatius Lawson, once a physician, now a coiner lately arrived in Newgate. He turned out to be perfect for the job. Most important, Newton owned him top to bottom. The Warden had several witnesses prepared to swear they had seen Lawson with coining tools, bending over a furnace, casting guineas and pistoles, trimming oversized pieces with metal scissors. Newton even had one man willing to swear that Lawson had said "he could make a fool of 20 such as the Warden."

Not anymore. Arrest and then weeks in Newgate had broken the former doctor. His co-conspirators had abandoned him, he told Newton, having "run away with all my goods starved one of my Children to death and sent the rest a begging." He was starving, trying "to live wth. a pennyworth of bread 4 dayes." He begged for help. "I throw myself wholly at yo[ur] feet," he wrote in his first letter to Newton, and in one of the last he pleaded once more: "I hope Charity will moe you to lend me yo[ur] hand . . . and the remaind[er] of my life I shall wholy rend[er] at your disposall."

Newton took Lawson up on his offer, and was repaid almost immediately. Lawson was a sponge, absorbing everything he heard in the cells. He did not confine himself to Chaloner alone. Among Lawson's early dispatches, Newton found the entire history of Ball and Whitfield's coining conspiracy, down to the damning tiny anecdote that "Ball sold his horse to a hatter in Southwark for to raise money to make ye pistols." A hatter! Now here was a witness to persuade a jury, ready to testify to the perfect detail that could convince twelve honest men he had actually seen what he presented as fact.

Lawson kept such tales coming. As he told them, coiners seemed driven to incriminate themselves just when he happened by. There was John Deacon, who in April 1698 approached him in the Swan tavern in the Leadenhall market to repair a guinea press. Katherine Coffee had tried to replicate one of Chaloner's coining techniques in "a Chamber at the sign of the Red Cow in makt Lane near St. James's makt" in front of six witnesses—including Lawson. "Perkins the Smith" had no more discretion: "one morning I came into his shop where he and his man Tom was striking with a Cast punch the face of a King William Shilling." For some reason, Perkins felt compelled to tell Lawson who had ordered the punches—a thief-taker named Wood—and Lawson dutifully passed that nut of information on to the voracious Newton.

Lawson went on, detailing incident after incident, recorded on page after page of Newton's files—clearly intent on delivering every last scrap he had heard, seen, or merely believed plausible (like the fashion detail that Katherine Coffee had carried her coining tools in "a black Leather bagg like the case of a small Bible"). But all this was garnish. Newton wanted the meat of the case, and that meant Lawson had to worm past Chaloner's sense of self-preservation and get the man to talk.

In this he had one great advantage over both of Newton's previous informers. Like Chaloner, Lawson had operated in the inner circles of London's coiners. He knew the same kinds of people—gold and silver dealers who supplied the raw materials, engravers and smiths who made and finished the coiner's tools, pub owners who provided meeting places and the occasional back room in which to work. But, crucially, he and Chaloner had never worked together. The cases against them did not overlap. Lawson could not testify against him. And so he slipped through Chaloner's defenses, becoming his intimate companion, eating, sleeping, and conversing in the same cell.

Chaloner seems to have welcomed the new man with real relief. Here at last was someone who couldn't hurt him, yet would recognize his mastery of their shared craft. He confided in Lawson, boasted to him, and accepted the sustained compliment that an eager listener pays his entertainer. Once he started to talk, all the secrets he had kept so carefully from such obvious spies as Carter and Whitfield came pouring out. After each long, wandering session Lawson passed on his report to the

Tower. By late January, Newton knew within a day—at most two or three—all that Chaloner feared, hoped, or planned.

Thus, when Chaloner wondered out loud if Carter had any Malt Lottery tickets left to produce against him at trial, Newton knew Chaloner was still preoccupied with the wrong charges. When Chaloner told Lawson that Patrick Coffee and Thomas Taylor, the goldsmith and engraver who had set him up as a coiner almost a decade ago, were at large, Newton knew which witnesses his prisoner feared most. Chaloner told Lawson that Katherine Coffee might be an equal threat, but thought she would "be ript up and dy before she confess any thing." This wasn't speculation; Chaloner had found a man outside the prison—a Mr. Hount or Hunt—to keep watch on her. Newton promptly took her deposition, in which she tied Chaloner directly to the making of forged French pistoles.

Chaloner told Lawson that he had much more to fear from Elizabeth Holloway—and, with two interviews already on record with her, Newton knew that he was right. Then there was Jack Gravener, brother to Joseph Gravener, who had married Chaloner's sister and worked with Coffee to gild Chaloner's first runs of pistoles and guineas. Joseph had already gone to the gallows, but Jack was still alive. Chaloner told Lawson that Jack could hang him, for he "hath seen him coyn many a thousand Gineas which he sold at 10s a piece."

Newton did not or could not track down the surviving Gravener, but Lawson's torrent of information kept coming. Chaloner boasted to Lawson that he had produced thirty thousand false guineas in all—which, even at its fifteen-thousand-pound street value, still amounted to a fortune, a couple of million in today's money. Chaloner took pride in his skills, preening to Lawson that "he graved the Plate for Mault Tickets and was to grave another for £100 tickets and could do a plate in 4 or 5 hours time & that no man in England could grave them better than he." Chaloner told his confidant some of his favorite tricks—how he used "buttons of Tin plated over with Silver," for example, frauds that could make a coiner a thousand pounds a week.

He also confessed to the petty crime of counterfeiting the three pounds that his agent Gillingham had used to cheat the captain carrying the Holloway children to Scotland—one of the deceits that had enraged Elizabeth Holloway to the point of telling Newton everything she knew.

All this was useful—vital, even. But Lawson truly earned his reward when he managed to extract from Chaloner exactly how he planned to derail his trial. By February, Chaloner had come to realize that whatever jeopardy he faced in the Malt Lottery scheme, Newton was clearly accumulating a much broader case. So he knew he had to find a way to render meaningless the growing mountain of testimony to his older crimes. Chaloner's first gambit was admirably direct. In his second report to Newton, apparently in early February, Lawson reported that Chaloner's "friends send him word this evening that they have made friends with 6 of one Jury and 8 of another to throw the bill out." (That is, he would suborn the grand jury to deny the bill of particulars—the indictment.)

At the same time, Chaloner developed a second line of defense. If he could not bribe his way out of Newgate, perhaps he could buy his escape with the one possession of real value he had left: the still missing Malt ticket plate. Secretary Vernon had already told him that "if he did not deliver up the plate . . . it should be worse for him." Accordingly, Chaloner was beginning to weigh the costs and benefits of so open an admission of guilt. He told Lawson that his treasure was in the custody of the daughter-in-law of the wife of an accomplice and indicated that he was wavering: "If he thought it would do him any Service he hath it ready to produce."

Weeks passed. Chaloner remained in Newgate, scheming. Lawson kept watch. Newton had already allowed two sessions of the criminal courts to pass since Chaloner's arrest. The next would begin on March 1, 1699.

Chaloner sensed that he was almost out of time. He could not be sure that his corrupted jurors would stay bought. He knew that he could not prevent at least some damaging witnesses from testifying in open court. He could think of only one last gambit to try, one unimaginable just weeks before: he decided to write to Isaac Newton and tell him why he should spare his life. It was the first time in the three years through which the men had fought that one of them had addressed the other directly.

Chaloner would tell all, he promised. "In obediance to your wor[shi]pp I will give you the best acc^t I cann remember." He would provide names, accomplices, men whose crimes he reckoned Newton knew, and "a great many more but I have not time to give you a whole acco^t of"—not with the days hastening toward the start of the next court session. But given space, hours, liberty, then "I should be glad to any Service to the Governm^t y^t is in my power."

23

———◦◦◦◦———

"If I Die I Am Murthered"

ISAAC NEWTON WAS PREPARED to listen—or rather, to take note of whatever Chaloner wished to tell him. Four letters survive in his records, three addressed to the Warden, one to a judge in the criminal courts which Newton copied into the file. One by one they register Chaloner's growing panic. Taken together, they form his last throw, an attempt through sheer weight of words to force Newton to accept what Chaloner wished to be true.

Chaloner began calmly. Not yet acknowledging his imminent peril, he sent Newton a two-paragraph note that, more in sorrow than in anger, claimed, "I am not guilty of any Crime." Oh, no: he was merely a bystander, caught up in others' wrongs. "Sr I presume you are satisfied what ill men Peers and ye Holloway are who wrongfuly brought me into a great deal of trouble to excuse their villainy."

Five years earlier, Chaloner had escaped punishment by sticking to his story that his former partner, Blackford, had betrayed him to save his own skin. Then as now, Chaloner's claim had the virtue of being mostly true. There was, of course, the problem that he had lately accused the Warden of incompetence and malice. Surely the Warden would not take a man's life over that regrettable incident. "You are greatly displeased with me ab^t the late bussines in Parliamt," Chaloner acknowledged. But the whole affair was, as usual, someone else's fault, as Chaloner had been compelled to appear by "Some p[e]rsons ag^t my desire."

Newton read this, and wrote . . . nothing.

Chaloner, unnerved, tried again, this time with more care to match his story to the evidence arrayed against him. So the Warden thought he knew the truth about Chaloner's early career? Not so. He had been no mastermind, just a go-between. His usefully dead brother-in-law Joseph Gravener had been the man in charge of a "great Trade in Clipping and Coyning." It was Gravener who rented a "Great Jewe's house in Mark Lane," fortifying it with "Iron barrs and very strong door on the staires strenghtend with iron" so that "20 men could not get in under an hour." There "they made ye Pistols some hundreds of them." Chaloner's role in this? "I knew all of these things his being my brother but I did not act." All he got was a stipend of forty shillings a week—walking-around money, a keep-quiet bribe.

Chaloner knew he needed to do more than put the blame for ancient criminal history on a dead man. There was Carter to account for, and behind him, the two-faced Davis. The Malt ticket plate itself, Chaloner said, was merely a bit of fun—a demonstration of his engraving skills, perhaps—"and if I intended to have anything to do in counterfeiting of Malt Tickets than I desire God Allmighty may never receive my soul." As for the people involved, wasn't it obvious that Davis had connived at the entire affair? "I understand by Carter," Chaloner wrote, "that Davis gave Carter a Ticket to do it by"—another strategically useful true statement. Then, when Davis "goes to the Governm^t and discovers that such a thing was going and got money to make further discovery," Carter took fright, and Chaloner destroyed the plate to ease his friend's terror. Thus, he concluded, "it may plane appear to an y Impartiall man that Davis made this Plott to get money out of the Governm^t."

And what of Carter's claim that Chaloner had orchestrated the plan and engraved the plate? It was all lies, false testimony, bought and paid for: "Davis comes to Newgate to Carter very often and bids him stand to w^h he has said, and said if we can

hang Chaloner I shall get 500£ . . . then I will get you out."
Could not Newton see the obvious, Chaloner asked. The wrong
man was in jeopardy: "I can planely make that these Tickets
bussines is a piece of roguery and forgery of Davis himself to
get money out of the Governmt."

At the last, this tone of injured innocence fractured. Panic
creeping in, Chaloner wrote, "I have been guilty of no Crime
these 6 years." He had been confounded by conspirators and
villainy; to prosecute him would be a sin: "if I die," he wrote, "I
am murthered."

Again, Isaac Newton did not choose to reply.

The silence harrowed Chaloner. If he could not get the Warden to respond, if he could not involve his accuser in his version of the story, his last weapon—his gift for persuasion—was
worthless.

He tried again. This time he wrote first to a magistrate in
his case, Mr. Justice Railton. In a long letter, he reminded the
judge of all the criminal deeds he had exposed over the past
several years. He had reported the Bank of England forgeries
that had prompted the Bank's board to take his advice (and
to seek, though not receive, a pardon for past crimes on Chaloner's behalf). There was his evidence about coiners operating
inside the Mint. And no one should forget the Jacobite printers
brought to the gallows on his testimony.

Now, he told Railton, he was paying the price for his services
to the Crown. "I was the cause of Carter's being put in the pillory before I discovered where Carter and his wife were coyning"—and now "their malice is so great that they endeav[ou]r
clandestinly to insinuate to the Governm^t y^t I would be concerned about Malt Tickets." He begged the judge to remember his services to the government. If he did, then neither "y^r
wor[shi]pp nor the Court will believe the suggestions of such
evill persons ag^t me."

No reply from Railton survives, probably because none was made: Chaloner's fate turned on the pleasure of the Warden of the Mint. And so the prisoner fell back on his final gambit. By the end of February, with his peril undeniable, Chaloner wrote to Newton twice more. The testimony against him was not merely false, he argued, but nonsense. He could not possibly have done the crimes of which he was accused. Why not? Because, he now said, he was incompetent, unable to perform the essential tasks of the coining trade. "I remember I have said to you that I understand graving," he wrote. "But whatsoer I have said no man in the world can say I can or could grave a flat stitch." (Flat stitch engraving involves the use of a flat tool with fine grooves cut in its head; it is used to inscribe parallel lines, hatching that adds depth to letters or crests. It demands considerable skill, and its use would have been essential to produce plausible copies of the Malt Lottery tickets.) "I can chase a little," he conceded, "but I never graved a letter of flatt stitch nor otherwise but to play the fool with a Graver as any man may do." Chaloner knew that there was a bit of a problem here. "I remember I said to you once yᵗ allthough I can grave and coyn," he acknowledged, but he had not intended that Newton take him literally. "I do not but it was but in generall speaking because I know the ways of graving Stamps from Taylor but pray examin Taylor if I can do any thing in flatt stitch."

This from the man who had just weeks before boasted to Lawson that there was nothing "to be p[er]formed in Coyn or paper but that he can effect with ease." Chaloner had always brandished his skills as a weapon. His mastery of the theory and practice of coining was the foundation of published claims that he could better the Warden. He had used his superior knowledge to force confederate after confederate to take the lion's share of the risk in every scheme. Even in Newgate, he had used his reputation as bait, trying to bend witnesses to his version of the story with both threats and the promise of riches to come—once the ever-elusive William Chaloner had again beaten the rap.

Now he disavowed it all, writing in his last letter before the trial that "I never could work at the Goldsmith's Trade in my life." He could not forge coins, his meager capacity sufficient only to run errands for men who could: "those pistolls was made by Coffee and Gravener and all that I ev[er] was guilty of was getting them the stamps." There was no cleverness in him, no nimbleness in his fingers. Again: "I never graved in flatt stitch graving in all my life and if now it would save my life I could not do it I have no Dye as God Allmighty shall judge me." And if once he had been tempted to sin, that impulse was no more: "The Tool I had before the parliam^t was wasted and spoiled long ago."

We know that Newton received this message, as he had the previous ones, for they survive in his Mint files. We can be virtually certain that he did not so much as contemplate a reply. When he did choose to answer a letter, he left traces, drafts and reworkings, four or more for sensitive documents in which he refined to perfection what was in his mind. Nothing of the sort exists in response to any of Chaloner's messages.

Chaloner understood the unbroken stillness of his enemy. Over his criminal career, Chaloner had spent a fair amount of time in jail—perhaps more than a year—but he had never actually come to trial. Now it was obvious that he must, and for a crime that could lead him to the hanging tree at Tyburn. At the very end, he broke. Lawson reported to Newton that Chaloner had gone mad, "pulling his shirt to pieces and running stark naked at midnight abot. the Ward for ½ an hour together."

The fits came and went, interspersed with periods of lucidity and milder delusions. In his first frenzy, Lawson wrote, "the men bound him hand and foot in his bed, but now he seems more rationall." In the restored quiet of the cell, Chaloner confided to Lawson the reason for his calm: "he hears very good newes yt they cannot bring andy Indictmt agt. Him and he questions not but [that he] slip out . . . as he hath done 5 times before."

Lawson seems to have believed that there was some truth in Chaloner's ravings. At least one other observer agreed. His biographer wrote that "the apprehension of what he might come to, struck him into a Fit of Sickness, and wrought so strong upon his Brain, that he was sometimes Delerious." When the madness came upon him "he was continually raving that the Devil was come for him and such frightful Whimseys."

At the same time, no one doubted that Chaloner recognized a good trick when he saw it—and he soon sought to put his madness to use. Lawson reported that he twice heard Chaloner say "that when the Sessions came and if he found himself in danger he would pretend himself sick otherwise he would be well and take his tryall." His biographer agreed. "These intervals of Lunacy," he wrote, "he endeavor'd to improve to a height sufficient to put off his approaching the Tryal, counterfeiting the Madman as well as he could."

24

∽◇∽

"A Plain and Honest Defence"

THE RUSE did not work. William Chaloner could not stop Newton's progress. The next court sessions opened at the beginning of March 1699 in the Guildhall, whose medieval Great Hall had served as the center of London's government since 1411. There, grand juries for London and the county of Middlesex convened. These panels were not mere rubber stamps, issuing indictments at the command of any ambitious prosecutor. Rather, they were designed to ensure that no one stood trial at the caprice of the sovereign or some powerful rival. Before he could proceed, each plaintiff was compelled to present enough of the evidence he planned to bring to the trial court to a jury of Englishmen who had the unchallengeable right to dismiss what they saw as frivolous or unfounded charges.

This was the hurdle Newton had failed to leap in his previous attempt to bring Chaloner to trial—and now, at last, he unwound the long game he had played to avoid a repetition. On March 2, as the Crown opened its presentation to the jury hearing cases that fell into the Middlesex jurisdiction, no mention was made of the Malt Lottery plot—unquestionably to Chaloner's surprise, given that he was still trying to disavow his engraving skill. Instead, Newton had prepared three completely different indictments.

To save some of his ammunition for the case he would present to the criminal jury, Newton used just two of the six witnesses he would call later, supporting their testimony with additional evidence developed over the course of his interrogations.

Those two witnesses—Thomas Taylor and Katherine Coffee —offered the jury what they claimed was direct knowledge of Chaloner's role in the matter of the forged French pistoles. Next, thanks to the information received from Elizabeth Holloway, Newton could provide sworn depositions to back up the story behind the next charge, that Chaloner had induced Thomas Holloway to escape to Scotland to undermine the earlier prosecution. And for the third indictment, again supported by his trove of depositions, Newton offered a story of Chaloner's forging a wide variety of English coins, from sixpences to guineas, in what amounted to a coining orgy. On a single day in August 1698, he charged, Chaloner had struck gold pistoles and guineas, along with almost one hundred silver pieces: twenty crowns, forty half-crowns, twenty shillings, and ten sixpences.

This last charge was absurd on its face. No coiner would have set up such a profligate and inefficient production line—six different coins, indiscriminately silver and gold. As Chaloner could have told him—as he had in fact written in his two published accounts of coiners' methods and tools—skilled workmen used molds and hammers or presses to make their coins. Each mold or die served a particular denomination. It would have hopelessly confused the process to change sizes, values, and metal recipes hour by hour across the day. Any sensible coiner set up the production line for one denomination and worked it until he or she was done. Newton surely knew this too, but he presented his story with a sufficiently straight face. Whatever corruption of the panel Chaloner had attempted, failed. The Middlesex grand jury for the March 1699 sessions of court returned three true bills against him, one for each of the crimes the Warden had alleged.

Chaloner, asked to plead to the indictments, stood mute. This was his last attempt to postpone the trial. English legal practice required an affirmative statement from the accused: guilty or not guilty. Standing mute dragged proceedings to a halt. There were, however, methods to persuade the obstinate. In the most gruesome, peine forte et dure, the silent accused would be taken to a cell and shackled to the floor. Warders would pile blocks of iron on the prisoner's body until he either pled or died. In Chaloner's case, two of the indictments could have earned that treatment, and the judges could have ruled his silence on the third as an admission of guilt. Chaloner bowed to the inevitable, and "at length he was prevail'd upon, and pleaded Not-Guilty."

Isaac Newton and William Chaloner fought their last battle the next day, March 3. English trials in the late seventeenth century were swift and ruthlessly to the point. There were no lawyers. Prosecutions in most felony cases were handled by the victims of crimes themselves, or by local authorities in cases, such as murder, where victims could not speak on their own behalf. Crimes against the Crown required some agent of the state—the Warden of the Mint or his designated mouthpiece, for example—to stand as the aggrieved party.

Chaloner had to speak for himself. There was no presumption of innocence. He had to offer an affirmative defense—either an outright argument of innocence or some demonstration that the prosecution's witnesses and evidence were sufficiently tainted so as to leave the case unproven. It was still an unpopular position that the defendants might benefit from the counsel of someone learned in the law. As the influential early-eighteenth-century legal scholar William Hawkins wrote, it should require "no manner of Skill to make a plain and honest Defence."

The trial took place in the Old Bailey, which stood just beyond the western side of the London city wall, about two hundred yards from St. Paul's Cathedral and conveniently close to Newgate. The building, erected in 1673 to replace the courts lost in the Great Fire of 1666, contained a ground-floor courtroom that was open to the sky, the better to reduce the risk that prisoners with typhus would infect judges and juries. (The danger was real. The courtroom was enclosed in 1737, and in the worst of the incidents that followed, sixty people died following a court session in 1750, among them the Lord Mayor of London.) Two upper stories loomed over the well of the court, leaving it in shadow for much of the day. The accused—Chaloner, when his day in court came—stood in the gloom on a platform, the infamous dock. There, standing behind the rail, the bar to which lawyers are still called, the prisoner faced his judges and the witness box in which he would by right confront those who would testify against him. Partitioned boxes to his left and right held the jury, and above the jury boxes, balconies on either side held respectable onlookers, peering down to complete the image of the courtroom floor as an arena, the den in which men and women faced the prospect of death.

Less worthy spectators packed into the yard behind the open end of the court. For many, the Old Bailey sessions provided a day out—a circus act—but the crowd also included (or so the authorities complained) criminals yet uncaught, preparing against the day they might find themselves facing judgment. Chaloner's entrance would have set an extra buzz running through that crowd, famous enough as he was to attract the seventeenth-century analogue of the celebrity reporter. One of those scribblers left a description of the Chaloner trial that is still the most vivid (if not entirely unbiased) portrait of Newton's antagonist.

When Chaloner's case was called, he had almost no time to think. The court sitting in the Old Bailey heard an average of fifteen to twenty cases a day; many took just minutes from start to finish. As the trial began, his predicament grew worse. In an age without advocates for the defense, there was a presumption that judges "should be the advocat for the prisoner in every way where justice would permit it."

Not this time. Standing alone at the bar, Chaloner stared at the formidably irascible Salathiel Lovell, Recorder of London—the chief jurist for the jurisdiction. Notoriously intemperate, Lovell had a reputation as a hanging judge. In one famous case involving a supporter of the deposed King James, he ignored the legal complexities his fellow judges had seen in another case and "cut asunder the Gordian knots of law he could not untie . . . and directed the jury to bring him in guilty, which they did." He had low friends too, conniving with the thief-takers who could be as much the source of as a solution to crime. Daniel Defoe, one of many who loathed Lovell, wrote, in *Reformation of Manners*:

> *Fraternities of Villains he maintains,*
> *Protects their Robberies, and shares the Gains;*
> *Who thieve with Toleration as a Trade,*
> *And then restore according as they paid.*

Worse still, according to Defoe, Lovell offered justice for hire:

> *Definitive in Law, without Appeal,*
> *but always serves the hand who pays him well:*
> *. . .*
> *He has his Publick Book of Rates to show*
> *Where every Rogue the Price of Life may know.*

Defoe was a master polemicist, and his claims are not facts. Absent hard evidence, the most that can be said is that if Lovell was not running a protection racket, he did turn a usefully blind eye in order to buff his reputation as a relentless investigator and scourge of crime.

All of which meant that in better times, a judge with Lovell's reputation would have been a perfect target for the cheerfully corrupting William Chaloner. Now, though, broke and unable to bribe anyone, much less as expensive a man as the Recorder, his only value to Lovell was as someone whose conviction could bolster the judge's credentials as London's chief crime fighter.

Chaloner was famous, with a trail of public bombast behind him to ensure that his conviction would be noticed in all the right places. He was friendless now as well, so Lovell need fear no covert attack on his own interests. Above all, the powerful —Newton, certainly, and Vernon, and behind them the ruling Whig establishment—wanted Chaloner gone. Lovell understood the value of pleasing those who could reward him. (Three years later, he would ask the King for a landed estate in recognition of his vigor in hounding coiners.) Chaloner could not have drawn a worse judge.

The Recorder made his influence felt from the beginning of the trial. Speaking from the bench, one of the judges—unspecified in the record, but almost certainly the vocal Lovell—opened the proceedings by calling the defendant notorious, clearly indicating to the jury which way the wind blew. The impression of an overwhelming presumption of guilt grew as Newton's parade of six prosecution witnesses entered the chamber.

With that entrance, Chaloner was able to gauge the direction of the testimony he would have to counter. Before he had an instant to gather his wits, however, the trial began.

That prosecution was a probably deliberate muddle. Newton seems to have taken to heart the advice he got a year earlier, that he could simply throw enough dirt around to convince the jury that Chaloner must have done something bad. The prosecution's witnesses essentially ignored the central claim of the indictment. Rather than dwell on proving that Chaloner had actually produced more than one hundred coins, both false gold and false silver, of five different sizes and designs, all in a single day, Newton's witnesses took the jury on an extended tour of the previous eight years of Chaloner's career.

So Thomas Taylor and Katherine Coffee repeated their story of Chaloner's early misdeeds. Coffee's story revealed that Chaloner had mastered the use of stamps and a hammer to make French pistoles by 1691. Carefully parsing her evidence, she reported that she had seen "Gineas which were reputed Chaloners but never saw him coyn any."

Taylor's information buttressed Coffee's. As Chaloner knew, staring across at his old supplier, Taylor could tell the court that he had provided two sets of stamps or dies—one for the pistoles Mrs. Coffee had just placed under the prisoner's hammer, and the other for English guineas. Never mind that the events in question took place seven years before the date of the crime for which Chaloner was supposed to be standing trial. Here he was in memory, caught red-handed committing treason.

The other four witnesses were sworn and spoke in quick succession. Elizabeth Holloway seems not to have testified to her Scottish odyssey, but she and Katherine Carter told what they knew—or what they were willing to say—about Chaloner's superlative skill as a coiner. Another witness agreed with Mrs. Carter that he had seen Chaloner make dud shillings on the day specified in the indictment. Both were almost certainly lying, at least in the details. In all the depositions Newton had taken over the preceding four months, several witnesses described Chaloner's June experiments with molded pewter shillings, but none mentioned any coining in August.

But even so, what could Chaloner say? It hardly advanced his cause to argue that he had made false coins two months before the day Newton's witnesses claimed, and that they were shoddy products, not the high-quality fakes they described.

The last to speak for the prosecution was John Abbot, the metal dealer turned coiner whom Thomas Carter had betrayed to Newton in January. Abbot swore that Chaloner approached him in 1693 or 1694, seeking to use his countinghouse. Abbot resisted, not wanting to turn Chaloner loose on his premises, "because wh silver and Gold he had was there." But eventually, as Abbot had told Newton, he cleared out a back room and locked Chaloner into that one space. On his return half an hour later, he opened the door of the countinghouse "& going in found the said Chaloner there in his shirt fileing of Gineas round ye edges and saw him edge them when he had done filing and that he edged them with a piece of Iron which had a grove running along the middle of it."

Moving on, Abbot reported that in 1695 Chaloner had shown him several blank stamps about the size of guinea dies—and which he said "he could get to be struck with the Tower Dyes and that they were fitted to be struck on both sides like a Ginea but broader and he said that Patrick Coffee could get them done any day at noon by a journeyman Smith in the Tower." Hearsay, surely, but it would have taken an exceptionally cautious jury not to be swayed by what Chaloner purportedly did next: he "told the Deponent that he had got his business done."

Here at last the muddle of the Tower dies was resolved—and Abbot was not done yet. He claimed that Chaloner had boasted to him of coining in a house in Mark Lane that yielded six hundred pounds' worth of half-crowns in nine weeks. He also testified that Chaloner had come to buy the silver he needed for the operation from Abbot's own shop, and had tried to pay his bill with counterfeit cash. When confronted with his dud coins, Abbot said, Chaloner first tried to brazen his way out of the debt by threatening to sue Abbot for failure to deliver promised goods. Abbot refused to budge, and Chaloner backed down, paying his reckoning with money that had in fact seen the inside of Newton's Mint. He then sealed the deal by giving Abbot a good dinner—"a Treat at the 3 Tuns in Woodstreet."

Chaloner could see the jurymen on either side of him. He could gauge the tenor of his judges. He must have understood what the prosecution was doing. Legal niceties be damned: his enemy had placed him at the center of enough crimes to hang him, even if they weren't the offenses of which he was accused.

The last witness answered a final question. The prosecution ceased. The judges faced the prisoner. What defense had he to offer? By Newton's careful design, Chaloner had almost no options left. He had not known which former friends would testify. He had no counsel, no legal advice. He had to speak then and there, with no chance to reflect, to organize an argument, to seek out witnesses of his own.

Even so, Chaloner was not completely without resource. He spoke out angrily that the court should recognize that the witnesses were perjurers, lying about his deeds to save their own guilty hides—a claim that was at least partly true. Chaloner was "very sawcy in Court, affronting Mr. Recorder [Lovell] divers Times," one observer wrote. But it was clear that neither judges nor jury were going to credit claims of perjury above his former associates' descriptions of specific crimes.

Chaloner retained one last hope. He had not been able to prepare for the testimony against him, but he had listened to every detail of it. He noted where he was supposed to have forged pistoles, crowns, half-crowns, and shillings. Abbot's shop—a London address. The Tower—within the precincts of the City of London. The Flask tavern—London again, and so on through each devastating retelling of offense after offense. And yet Chaloner faced charges brought by a Middlesex grand jury, being heard by a Middlesex trial jury. How could such a court, Chaloner asked, address crimes committed outside its jurisdiction?

It was a neat argument, and in fact the law was on Chaloner's side. Both the Middlesex and the London grand juries met in the same hall at the start of each criminal court session, and both returned indictments to be heard in the Old Bailey. On one day for which a detailed schedule survives, the court opened with two London cases heard before a London jury, followed by the trials of eight Middlesex offenders, all before the correct Middlesex jury. Those ten trials finished by the midday meal break. Such chopping and changing happened all the time—and it reflected the problem of matching legal tradition to the sprawl of the metropolis, which created a continuous habitat for crime that extended far beyond the formal boundaries of the old City of London.

So Chaloner's was not a purely random shot. A similar strategy would work for his nemesis John Ignatius Lawson when he came before a Middlesex jury later that year. Although he had already confessed to crimes sufficient to hang a dozen men, those specified in the indictment—charges that Newton would have had to prepare—had been committed in London, and thus, the judges in that case ruled, could not be addressed in the Middlesex court. In what can only be seen as Newton's payment for services rendered, Lawson walked out of court a free man.

A sympathetic judge could have similarly ordered Chaloner's release. A prosecutor who merely wished to put a scare into a potentially useful prisoner could have eased the court to that conclusion. But not this time. There was no one in court with any charity toward William Chaloner. Lovell and the rest of the judges ignored his objection.

The rush of the court, the pressure of the dozen other cases to come that same day, meant he had to speak and be done—and he did, having offered what observers judged "but an Indifferent Defence." "The Evidence was plain and positive," so after Chaloner had spoken, the judges gave the case to the jury. Juries in this period would retreat to a separate chamber only in complicated cases where they had to decide on the guilt or innocence of several defendants. For a simple matter, they would huddle together for a minute or two in the middle of the courtroom.

They did not keep Chaloner waiting long. Ignoring the lesser charges, the jury "soon after brought him Guilty of High-Treason."

The next day, March 4, 1699, the prisoner stood once more at the bar of the Old Bailey, now to hear his sentence:

Death, by hanging.

The trial of William Chaloner was over.

25

"O I Hope God Will Move Yo᷎ Heart"

CHALONER DID NOT GO silently to his appointed fate. "After his Condemnation, he was continually crying out that they had Murder'd him," his biographer recorded, that "the Witnesses were perjur'd and he had not Justice done him." He fought, he raged, he whimpered. Following the conventional demands of the true-crime genre, his one biographer mocked Chaloner's terror, writing that he "struggl'd and flounc'd about for Life, like a Whale struck with a Harping Iron."

One hope remained to him. The presiding judge had to forward the death sentences imposed at each court session to the King and his ministers for review. The court could recommend clemency—but not Lovell's, not in this case. On March 19, Secretary of State Vernon brought nine capital cases from the sessions to William III. Two of the condemned received royal mercy, and lived. But Chaloner "was too well known to be credited"—certainly to the men handling the appeal. In essence, whatever the defects of his trial, Chaloner's crimes were too familiar to those in the know. Thus "his Character contributed to his Ruine . . . so that the Warrant for Execution being sign'd he was amongst the number appointed to die."

Chaloner received the news in Newgate. He was still being watched, more for entertainment now than for any judicial purpose. Thomas Carter reported to Newton that "Chaloner . . . p[er]sistted to the last how in[no]sent he was for wh. he dyed." His biographer added (or, as likely, manufactured) the sensational details: on hearing the King had signed his death warrant, Chaloner "bellow'd and roar'd worse than an Irish woman at a Funeral; nothing but Murder! Murder! Oh I am murder'd! was to be heard from him." He was inconsolable: "nothing cou'd be thought on to make him take that patiently, which he must embrace whether he would or no."

Chaloner was certainly terrified. In a last letter to Newton, he started off badly, writing as if there were still something to be argued: "allthough p[er]haps you may think not but tis true I shall be murdered the worst of all murders that is in the face of Justice unless I am rescued by yor mercifull hands." He reprised all the defects of his "unprecendented Tryall": that none of the witnesses told the court they had actually seen him coin; that London crimes could not be tried by Middlesex juries; that most of the testimony did not bear on the date specified in the indictment; that the witnesses perjured themselves out of malice and self-interest.

Toward the end of the letter he seems to have realized that his tone was hardly likely to persuade the man who had orchestrated every detail of the proceedings that had brought him to the edge of disaster, and in his final passage Chaloner abandoned any semblance of argument. "My offending you has brought this uponn me," he wrote. But could not his enemy relent? "Dear S[i]r do this mercifull deed O for God's Sake if not mine keep me from being murdered."

And then: "O dear S[i]r nobody can save me but you O God my God I shall be murderd unless you save me O I hope God will move yor heart with mercy and pitty to do this thing for me."

And once more:

I am
Yo[u]r near murderd humble Servt
W. Chaloner.

Isaac Newton, victorious at last, did not trouble himself to reply.

The morning of March 22 found William Chaloner in full cry. A day or two earlier, in a last gesture of bravado, he had sent the long-missing Malt copper plate to the Tower—a gift for the Warden of the Mint. But now, when his jailers came for him, he brandished a list of complaints and demanded that it be printed. He was refused.

To the chapel next, where he joined the other prisoners bound for the gallows. He may have sat before the coffin that was sometimes placed on a table before the condemned men's pew. When the chaplain urged him to show the proper spirit of repentance, Chaloner refused, shouting with "more Passion than Piety." The chaplain tried to calm him, but Chaloner raged on. "Notwithstanding the great Care and Pains of the Reverend Ordinary, twas difficult to bring him to a sense of that Charity and Forgiveness proper to all Christians, but more especially to [d]ying Men." Finally, Chaloner steadied himself enough to receive the sacrament, and the doomed worshipers filed out into the open air.

The convoy set out at about noon, bound for the traditional execution ground at Tyburn, now Marble Arch. Some of the condemned men traveled in style. John Arthur, an infamous highwayman, sat at ease in a coach and was cheered by the crowd as he paused at public houses along the way, arriving at the gallows as drunk as he cared to be.

Chaloner had no such comfort. Once Parliament turned coining into a species of high treason, the execution of coiners followed the same brutal sequence of punishment laid down for those found guilty in Guy Fawkes's Gunpowder Plot. A traitor drank no gin. No one cheered his name. Chaloner was brought to the place of execution on a rough sledge—no wheels. There were no underground sewers in seventeenth-century London, only courses in the roadways to carry sewage to the river. As the sledge bounced along, fountains of filth would have erupted, human and animal waste splattering his clothes, arms, face. All the while he continued to call out his innocence, crying "to the Spectators that he was Murder'd by Perjury." He would have reached the execution ground at Tyburn stinking, wet, cold, and mercilessly sober.

The method of execution for traitors had been in place since Edward I killed the Scots insurgent William Wallace. The condemned must be "hanged by the neck but not until you are dead . . . taken down again, and that whilst you are yet alive, your bowels [must] be taken out and burnt before your faces, and that your bodies be divided each into four quarters, and your heads and quarters be at the King's disposal." Counterfeiters got a reprieve: they were permitted to choke on the noose till they died, so that any mutilation of their bodies would take place on their corpses.

At his turn for the gallows, Chaloner once more cried out that "he was murder'd . . . under pretence of Law." A minister approached him and again bade him show the penitence and forgiveness demanded of those about to die. This time Chaloner accepted his set role and paused for a moment "to pray with much fervency."

The rope dangled from three crossbeams set in a triangle —Tyburn's Hanging Tree. Prisoners mounted a ladder to put their heads through the noose. Trap-door gallows that could kill quickly would not enter common use in England for another sixty years. When the moment came, the executioner's men would pull the ladder out of the way and the condemned dangled, twitching and jumping (the "hangman's dance") as long as it took—sometimes several minutes—for life to choke out of him.

Chaloner showed courage at the end. He mounted the ladder. Then, "Pulling his Cap himself over his Eyes, [he] submitted to the stroke of justice." Richer men often paid the hangman to pull on their legs to speed death. Not the destitute Chaloner. He had to choke till he drooped, to the greater amusement of the crowd.

William Chaloner lies in no known grave. He does possess an epitaph, the last lines of the biography printed within days of his execution:

"Thus liv'd and thus dy'd a Man who had he square'd his Talent by the Rules of Justice and integrity might have been useful to the Commonwealth: But as he follow'd only the Dictates of Vice, was as a rotten Member cut off."

"He Could Not Calculate
the Madness of the People"

ISAAC NEWTON did not attend William Chaloner's execution—there is no hint that he even considered doing so. He had other counterfeiters to pursue, and his work continued in the narrow rooms of the Tower.

Apparently, though, criminal London occupied him less once Chaloner was gone. After compiling more than two hundred interrogations through the peak months of that investigation, Newton collected just sixty more on various cases over the next year and a half. He may simply have been resting on the laurels he had earned for his work on the recoinage. From a purely manufacturing perspective, England's currency was sounder than it had ever been. According to the Mint's teller, Hopton Haynes, under "this gentleman's care we have seen it brought to that extraordinary nicety . . . as was never known in any reign before this."

Results like these earned Newton his reward at the end of 1699. Thomas Neale, the impressively useless Master of the Mint, displayed a previously unobserved knack for timing by dying in December. Within the web of patronage and obligation that ruled English politics, the mastership was a lucrative plum. Apart from a salary of £500 a year, the Master received a fee for every pound of metal coined at the Mint. Neale had made an additional £22,000 that way during the recoinage, even though Newton had done the work. Newton was not a particular political favorite; nonetheless he became the only Warden in the history of the office to move directly into the Master's position, in what was clearly a response to his role in saving England's coinage. He took up the new post on his fifty-seventh birthday, Christmas Day, 1699.

With that his fortune was made. The Mint remained busy enough to throw off extraordinary sums from time to time. In his first year, Newton took in £3,500 — enough to persuade him, finally, to give up his no-show Cambridge professorship, with its paltry stipend of £100. He would have much less profitable periods, but according to a calculation by Richard Westfall, Newton on average took in about £1,650 a year during his twenty-seven years as Master. Never truly poor, he was now well on his way to being genuinely rich.

In the new century, Newton also returned to some of the questions in natural philosophy that had consumed him in his youth. At the end of 1703 he became president of the Royal Society after another death, that of his old antagonist Robert Hooke, and within a couple of months he presented the society with the manuscript of the second of his two great books, *Opticks*.

Opticks reported the results of the investigations of light and color that had first brought Newton to the attention of the Royal Society back in the early 1670s. The book also presented Newton's first full declaration of all that he believed to be true, across the range of investigations that had consumed his life. He argued for intellectual humility; in a draft of the introduction he acknowledged, "To explain all nature is too difficult a task for any one man or even for any one age. Tis much better to do a little with certainty & leave the rest for others that come after than to explain all things by conjecture without making sure of any thing." But he maintained the unity of natural phenomena, writing about the concept of forces acting on bodies from a distance: "It's well known that Bodies act upon one another by the Attractions of Gravity, Magnetism and Electricity; and these Instances shew the Tenor and Course of Nature," for, as Newton argued, in one of his most famous epigrams, "Nature is very constant and conformable to her self." Thus, he argued, other such hidden forces would be found throughout the natural world.

Perhaps most significantly, he allowed himself to declare publicly a private conclusion he had reached long before. He acknowledged that mechanical ideas tended to eliminate the necessity of God: "Latter Philosophers banish the Consideration of such a Cause out of natural Philosophy," he wrote, "feigning Hypotheses for explaining all thing mechanically and referring other Causes to Metaphysicks." But this was an error in method, he asserted, announcing that for his part he sought to deduce "Causes from Effects till we come to the very first Cause." Not for him merely "to unfold the Mechanism of the World," but to learn "Whence is it that Nature doth nothing in vain; and whence arises all that Order and Beauty which we see in the World."

Newton knew the answer. "Is not infinite Space the Sensorium of a Being incorporeal, living and intelligent, who sees the things themselves intimately, and thoroughly perceives them, and comprehends them wholly by their immediate presence to himself . . . ?" Thirty-five years before, Newton had contemplated abandoning Cambridge because he could not swear to the claims of the Anglican Church. That was his response, a credo that he could affirm without reservation.

For all the grandeur of vision expressed in *Opticks*, however, the science it contained was old. The most recent experiments it reported dated back twenty years; most had been completed a decade or more before that. By the early 1700s, and perhaps a decade earlier, Newton mostly ceased to be a natural philosopher. In his remaining years, he focused on historical and religious approaches to a fuller knowledge of God. He contemplated the true nature of Christ's body; he speculated about the lives of God's agents on earth (among whom he numbered himself) after the Apocalypse; using the Bible, he tried to calculate the end of days—he reckoned that the second coming would not occur before 2060. Though some of his work on these questions was published posthumously, while he lived this was a private passion. Convinced though he was that his science, his mathematics, and his historical research tended to the same ultimate truth, he believed his conclusions to be "strong meats for men," and thus, as was his lifelong habit, he kept his boldest thoughts to himself.

Even so, Newton continued to play an important public role. The Mint still demanded a surprising amount of his time and effort, especially once the fate of the Great Recoinage became clear. As he had predicted, the recoinage, however successful as an industrial operation, was a failure as monetary policy. The decision to recoin without devaluing had the predicted result: silver continued to flow across the English Channel, buying continental gold at cheaper prices than those offered by the exchange rate between silver shillings and golden guineas. By 1715, most of the new silver specie struck through 1699 had vanished. In response, more or less by accident, the basis of British currency shifted from silver to a new gold standard.

Newton, first of necessity and then with more intention, oversaw this shift. In doing so, he found himself exploring the same kind of global information networks he had used to advance his arguments in the *Principia*. This time, instead of data on the tides and observations of comets and the motion of pendulums at various spots on the planet, he investigated what he quickly realized was a worldwide trade in gold. By 1717, he was able to sum up in detail what was going on. Gold was much cheaper in China and India than in Europe, Newton told the Treasury. That imbalance sucked silver—much of it mined originally in the New World—not just out of England but out of the entire

European continent. This was its own kind of action at a distance: the faraway, almost occult attraction of Asian gold markets putting European silver into a predictable trajectory, one to be explained with the same habits of mind that had brought the revolutionary study of gravity to its completion thirty years earlier.

At the same time, Newton remained alert to the limitations of seeing metal as the only true money. He defended the idea that paper, government borrowing at interest, could repair the deficiencies of an all-metal money supply. In essence, he argued for an inflationary policy, suggesting that the various experiments in debt of the previous decade—Malt Lottery tickets, Bank of England notes, Exchequer bills, and the rest—were practical and prudent answers to shortfalls in hard money. In a strikingly modern-sounding passage, he wrote, "If interest be not yet low enough for the advantage of trade and designs of setting the poor on work . . . the only proper way to lower it is more paper credit till by trading and business we can get more money." And even more radically he wrote, "Tis mere opinion that sets a value upon [metal] money," adding, "We value it because we can purchase all sorts of commodities and the same opinion sets a like value upon paper security."

In this Newton's was the minority view; even his old currency ally Lowndes disagreed with such an accommodating idea of the role of credit in the currency. Newton was right, though, and his notion closely approximates the modern conception of money. But these new conceptions of money were still poorly understood, even by someone with his analytical powers. In the last decade of his life, Newton would find out just how easy it was to be overcome by the promises written on bits of fancy paper.

It had seemed a good idea at the time. In 1711, English speculators formed the South Sea Company to take advantage of opportunities created by the War of Spanish Succession. They intended to exploit a government-granted monopoly trade with Spain's Latin American colonies, at a time when Spain could not enforce its own control of commerce there. The price paid for the monopoly: the South Sea Company agreed to take over some of Britain's official debt—that whole bestiary of obligations, bonds, and lotteries issued to pay for the nation's wars. The company recapitalized the debt with a loan of £2.5 million to the government and then converted the older obligations it had received into shares in the new company.

The promised trade never materialized, and the company began to act almost exclusively as a kind of bank—and an innovative one at that. In 1719, Parliament passed a bill permitting the South Sea Company to purchase more government obligations, again converting a range of public debts into a single, easily tradable form: stock in a company that could be bought and sold on the nascent market in London's Exchange Alley.

The creation of a permanent, easily transferable debt would prove to be a very valuable tool, one that some historians have credited with financing the great leap to global power the British Empire achieved over the next century and a half. But that financial revolution did not occur without the occasional setback—including, notably, the fiasco of the South Sea Bubble.

The bubble began with rumors circulating in January 1720 —set in motion by company insiders in a game as old as markets —that the trading side of the South Sea enterprise was about to take off. Exchange Alley bit, hard. South Sea stock rose from £128 to £175 a share within a month, and then the announcement of a new deal for the company to take on yet more national debt kicked the price up to £330 by the end of March.

That was just the beginning. The sense that there was easy money to be found fueled a speculative boom. By May, the price of South Sea stock topped £550, and just a month later, shares in the company peaked at £1,050, propelled to those heights by the announcement of a ten percent dividend, to be paid in midsummer.

Then it fell apart, fast. However it had begun, by the end the South Sea Company had devolved into a pyramid scheme, the classic con in which money from the last investors goes to pay off earlier punters with rewards that seem—and are—too good to be true. Eventually, all such schemes run out of new takers, and they collapse. Shares in the company started to fall in July, although in August they still commanded as much as £800. Then the bottom fell out. The stock price crashed to £175 within a month, wiping out virtually all those investors who had leaped, just weeks before, onto what had seemed an infallible money-making machine.

Among those last-in, first-crushed losers: Isaac Newton. He had actually been one of the early, and hence in theory, least vulnerable investors in the company. He listed a substantial amount of South Sea stock among his holdings as early as 1713, and he had sense enough to sell some of his shares into the rising market of April 1720. But the stock continued to rise, and Newton, watching as bolder players held on for a further threefold gain—on paper—succumbed a second time. In June, at the very peak of the boom, he directed his agent to purchase an additional £1,000 of stock. He bought more shares a month later, just as the price was beginning its slide. When the crash came, his niece, Catherine Conduitt, reported that his losses topped £20,000, roughly forty years of his base salary as Master of the Mint.

Newton, of all people, should have been able to penetrate the flaw in the math behind the South Sea fraud, the same that lies at the heart of every pyramid scheme. Look at the promised payments over time, expand the series—the very type of problem Newton first solved in 1665—and in short order the sums on offer exceed the total available store of money to pay them. Yet people who are offered a gold-plated promise of twenty percent or better returns on their money leap for the prize again and again. Newton did too.

The loss undoubtedly hurt, though Newton had not gone so far as to bet all he had on the bubble. He continued to be one of the largest individual owners of East India Company stock, with £11,000 invested in that much more stable business as of 1724, and the value of his estate as calculated a few years later topped £32,000, excluding his landholdings in Lincolnshire. So by any measure he remained a wealthy man. But the memory of the disaster pained him, and it was said he hated it when anyone so much as mentioned the South Sea Company in his hearing. It may not have been just the money lost that irked him so. Rather, it also seems that he saw he had been played for a sucker, like any mere unphilosophical fool. Once, speaking of the spellbinding rise in South Sea shares at the peak of the mania, he told Lord Radnor "that he could not calculate the madness of the people."

Whatever Newton's regrets, his friends remembered him in his last years as generally content, a more benign figure than the ferocious intellectual infighter of his earlier years. Despite his wealth, he lived moderately: bread and butter for breakfast, wine usually only at dinner. According to his niece, he hated cruelty to animals. He was genial to old friends, and despite his history of aloof unsociability, he became something of a pater-familias to his extended family. He was a fixture at weddings, where "he would on those occasions lay aside gravity, be free, pleasant, and unbended." Even better, from the family's point of view, "He generally made a present of £100 to the females and set up the men to trade and business."

As Newton passed into his eighties, the pace of his public life slowed. He ceased to take much active interest in the Royal Society, and some of his comments there betray a man more lost in memory than caught up in current intellectual concerns. The Mint he mostly left to its own devices, ultimately passing its management on to his niece's husband, John Conduitt, who succeeded him as Master. From 1722, his health began to decline. Gout and a nasty respiratory illness were enough to persuade him to move to Kensington in 1725, then considered to be "a little way of in the country," with sweeter air than the fug of London proper. Through that year and the next, he continued to read and write and think, his studies still centered almost exclusively on biblical history.

In February 1727, a visitor came to call. He found Newton attempting to ready his *Chronology of Ancient Kingdoms* for the printer. The old man entertained his guest by reading from the manuscript until dinnertime. A few days later, Newton attended a meeting of the Royal Society. The next day, he had an excruciating pain in his abdomen, caused by what was diagnosed as a stone in the bladder. The illness persisted for almost two weeks, and then, briefly, he felt the worst of his suffering abate. This hint of recovery was an illusion. He lost consciousness on March 19, and died in the early hours of the twentieth. At the last, Isaac Newton refused to take communion in the Church of England.

Before his death, Newton offered his own version of an epitaph. In perhaps his most famous moment of self-reflection, he wrote:

> *I don't know what I may seem to the world, but as to myself, I seem to have been only like a boy playing on the sea shore, and diverting myself in now and then finding a smoother pebble or a prettier shell than ordinary, whilst the great ocean of truth lay all undiscovered around me.*

Those who had known him took a different view. In 1730, John Conduitt was considering the design of Newton's monument in Westminster Abbey. He received a letter from a man who had once engaged Newton's thoughts as deeply as anyone ever would. Nicholas Fatio de Duillier remembered when the *Principia* had appeared like prophecy, a revelation. Thus he proposed the text for the inscription to be carved into the memorial: "*Nam hominem eum fuisse, si dubites, hocce testatur marmor.*" The phrase can be translated, "If you doubt there was such a man, this monument bears witness."

Acknowledgments

This book has depended on the kindness—and much more—of a host of people. Three of them are tied for first among equals: my editors Rebecca Saletan (Houghton Mifflin Harcourt) and Neil Belton (Faber and Faber), and my agent, Theresa Park. There is no gift I could have received to match the sustained, critical attention that I have received from Becky and Neil. Theresa's involvement in this book extends to its prehistory, and her kind and implacable guidance has been invaluable throughout the project. My thanks also to Deanne Urmy, who guided the book through its final preparation for publication.

I must also thank Bantam's Ann Harris, editor of my previous book, *Einstein in Berlin*, who extended the extraordinary education she gave me in the art of writing on that book by working through the ideas that became this one. Every writer should have such a generous sounding board.

Etienne Benson—now Dr. Benson, then a graduate student at MIT—was an invaluable research assistant, smart, fast, and insightful. My thanks as well to Houghton's Larry Cooper, whose manuscript editing was both rigorous and humane; the readers of this book may not know they owe him a debt of gratitude, but they do. Becky's assistant editor, Thomas Bouman, ably helped in the editing of the book, and only those who have done it themselves know just how much I owe him for his heroic work turning my footnotes into a publishable resource.

I have also benefited greatly from the help of the Newton scholarly community, fabulously learned and a group unusual, in my experience, in their generous welcome to a newcomer. Cambridge University's Simon Schaffer has a global network of students and colleagues who have benefited from his seemingly inexhaustible insights into Newton's work and times; I am just the latest who needs to thank him for early advice and the review of several versions of the manuscript.

Jan Golinski of the University of New Hampshire and Mark Goldie of Cambridge University did the same—several meetings, reviews of the manuscript, advice and encouragement. Many from the broader community in the history of science and economics also gave me invaluable help. Peter Galison at Harvard University gave me early advice and reviewed a late draft. David Bodanis, science writer and public intellectual extraordinaire, gave the manuscript a close read and a very sensitive critique. My MIT colleagues Peter Temin and Ann McCants, and my friend from across the river, Boston University's Zvi Bodie, reviewed the economic history sections of the book and improved the arguments there immensely.

Physicists Sean Carroll and Lisa Randall sought to straighten out any kinks in my explication of Newton's physics. Matthew Strassler listened to a lot of attempts to make sense of the roots of his science. Hilary Putnam once again gave me the enormous compliment of his unique combination of formidable learning and meticulous attention as he listened to my accounts of Newton. It should go without saying, but I'll say it anyway, that any errors of fact or interpretation that remain in this book are mine alone.

My thanks as well to Sallie Dixon-Smith of the Tower of London's curatorial staff, who very kindly led me through the former Mint precincts at the Tower, and Peter Jones of the Kings College Library at Cambridge, who gave me early leads on certain Newton manuscripts. David Newton, one of Isaac's distant relations, found documents relating to Chaloner's trial at the London Archives under considerable time pressure.

Early in the project, Rob Iliffe, now of Sussex University, Scott Mandelbrote of Cambridge University, Mordechai Feingold and Jed Buchwald, both of Caltech, and Owen Gingrich of Harvard University all gave me help and direction. Anne Harrington and her colleagues in Harvard's history of science department gave me a research home at a critical point.

I owe a special debt to the staff of Harvard's Widener Library, in which much of this book was written. My thanks also to those manning the desks at the British National Archives at Kew, where I was able to review all of Newton's Mint papers; at Newton's birthplace, Woolsthorpe Manor, where I received an out-of-season tour; and at the British Library's rare book and music reading rooms. History is a collaborative passion, and I could not have indulged my compulsion to engage the past without the generosity of so many others who share the same desire.

In addition to those acknowledged above, there are several people I never got the chance to meet whose work deeply influenced this project. I would not have known how to begin without Frank Manuel's attempt to understand Newton's emotional life in his *Portrait of Isaac Newton*. Robert Westfall created a body of scholarship that every subsequent writer on Newton has mined. I have too; his *Never at Rest* remains the definitive comprehensive biography (at least in English). I did meet I. Bernard Cohen—I even took one graduate seminar with him almost three decades ago. If I'd known then what I know now about the depth to which he followed Newton's thinking, I would have been able to thank him properly for all I have learned from his work. Last, I want to draw special attention to Betty Jo Teeter Dobbs, who did so much to rehabilitate Newton's alchemy as an integral part of the totality of Newton's thought, science, faith, and motivation. She did so in the face of scholarly opposition, and she won her basic point by dint of persistent, elegant, and ferociously smart intellectual labor. The study of Newton's alchemy has since been taken up by many others, but she was one of the pioneers, and I am in her debt.

I owe a special debt of thanks to my MIT colleagues and students. Marcia Bartusiak, Robert Kanigel, Alan Lightman, and

Boyce Rensberger of the MIT Graduate Program in Science Writing gave me help, timely advice, and the useful knowledge that books do finally get written, no matter how many papers must be graded in the meantime. My colleagues in the Program in Writing and Humanistic Studies were equally supportive, and I particularly valued conversations with Professors James Paradis, Kenneth Manning, and Junot Díaz, along with our visiting colleague Ralph Lombreglia, at various points in the maturation of the book. The deans of the School of Humanities, Arts, and Social Sciences, Philip Khoury and Deborah Fitzgerald, provided generous research support, and the dean of the School of Science, Marc Kastner, added valuable encouragement. My thanks as well to Rosalind Williams of MIT's Science, Technology, and Society Program, and John Durant, director of the MIT Museum. Science writer Gary Taubes has been a writer's escape hatch when the process closed in, and Jennifer Ouellette has also given gratefully received advice and counsel late in the game.

Family and friends are the safety net without which I could not attempt the high-wire act of writing any book. Hilary and Ruth Anna Putnam; Robert, Toni, and Matthew Strassler; Theo Theoharis; Michael, Isabel, and Thomas Pinto-Franco; Eleanor Powers; Lucinda Montefiore and Robert Dye; Simon Sebag-Montefiore; Geoffrey Gestetner; David and Juliet Sebag-Montefiore; and Alan and Caroline Rafael all lent ears, and sometimes beds, well past the point where the words "Isaac Newton" can have had any freshness to them. I am a lucky man to have such people in my life. My uncle Dan died just as I was writing these acknowledgments. He and my aunt Helen have helped keep me sane through all four of my books, and I cannot say how much I regret that Dan won't be able to help me see this one through. My siblings, Richard, Irene, and Leo, and their spouses and children, Jan and Rebecca, Joe, Max, Emily, and Eva, found the perfect balance: never (well, hardly ever) asking how the book was going while giving every appearance of enjoyment as I told them each newly uncovered tale.

Last and first, my wife, Katha, and my son, Henry, are the constant joys of my life. They gave me support, time, shoves when I needed them, hands-up as appropriate, laughter, and crucial perspective on what is, after all, a very odd way to make a living.

This book would not be here were it not for them. I cannot thank them enough, but I can try.

A Note on Dates

England during Isaac Newton's life used the Julian, or Old Style, calendar. At that time, the Gregorian, or New Style, calendar—the one we use today—had already been adopted on the European continent. The calendars differed in two important ways.

When the Gregorian calendar was adopted (the 1580s for most of Catholic Europe), the Julian calendar was ten days out of alignment with what was presumed to be its original starting point, to which the Gregorian returned. By Newton's birthday, December 25, 1642, the error had grown to eleven days, making January 4, 1643, his birth date in the new calendar. The other difference between the English calendar in Newton's time and current practice lay in the start of each year. January 1 did mark the traditional celebration of the New Year festival, but the legal year began on March 25. Dates between those two markers were often written in the form "January 25, 1661/2."

In this book I have used the dates as Newton would have known them—that is, following the Julian calendar as it was used in his time—with one exception: I turn the year on January 1 and use a single number to mark the passage of time.

Notes

⁘

PREFACE: "LET NEWTON BE"

PAGE

x "such vacan[t] places": "John Whitfield's Lettr to the Isaac Newton Esqr Warden of His Maj^tys Mint Febry 9th 98/9," Mint 17, document 134. Chaloner—or any counterfeiter: Only men were supposed to be hanged for currency crimes. Convicted women faced a worse penalty: they were to be burned to death. This punishment was rarely carried out by the late seventeenth century. See V.A.C. Gatrell, *The Hanging Tree,* p. 317.

xi a bust of Newton: Mordechai Feingold drew my attention to this painting in his *The Newtonian Moment,* p. 180. Feingold's book provides a wealth of detailed insight into the immediate reaction to Newton in both learned and popular culture, and his chapter seven, "Apotheosis," offers a valuable account of the myth-making that followed Newton's death.

xii four million pounds: Assessing monetary value across three centuries is a difficult and highly inexact process. But estimates of purchasing power, though imperfect, do confirm that Chaloner, if he was telling the truth, had been prodigiously successful. Among the most rigorous estimates comes from a research paper, published in 2002 by the Library of the House of Commons, that provides an index of the value of the pound from 1750 to 2002. In that calculation, one pound at the beginning of that period would be worth just over 140 pounds in 2002 (Library of the House of Commons, "Inflation: The Value of the Pound, 1750–2002," Research Paper 02/82, 11 November 2003). At that conversion rate, Chaloner would have produced about 4.2 million pounds of false currency in his eight years or so of coining. The number is probably a little lower than that, though again, any definite statement has to be hedged with fudge factors, acknowledging the drastic difference in patterns of consumption between the two ages. But E. H. Phelps-Brown and Sheila V. Hopkins laboriously constructed a record of the value of a building craftsman's wages dating back to the thirteenth century, and their analysis shows that prices were higher in the late 1690s than in 1750. ("Seven Centuries of the Prices of Consumables Compared with Builders' Wage-Rates." *Economica* 23, no. 92, new series, November 1956, pp. 296–314). If we apply a slight correction, it is fair to say that Chaloner's truly self-made fortune totaled somewhere between 3 and 4 million pounds in today's money. In other words, a lot.

1. "EXCEPT GOD"

3 carriage ride to the college: Isaac Newton, Trinity Notebook, Cambridge Ms. R. 4. 48c., f. 3.

admitted into its company: Richard Westfall, *Never at Rest*, pp. 1, 66.

4 dependence on another human being: This summary of Newton's birth and early raising is based on the account in Richard Westfall, *Never at Rest*, pp. 44–53. Westfall's account is largely derived from C. W. Foster's article "Sir Isaac Newton's Family," *Reports and Papers of the Architectural Societies of the County of Lincoln, County of York, Archdeaconries of Northampton and Oakham and County of Leicester* 39, part 1, 1928.

"outstrip them when he pleas'd": William Stukeley, *Stukeley's memoir of Newton in four installments*, Keynes Ms. 136.03, sheet 4.

5 portraits of King Charles I and John Donne: John Conduitt, Keynes Ms. 130.3, 12v and 13r.

"Isaac's dials": William Stukeley, *Memoirs of Sir Isaac Newton's Life*, Royal Society Ms. 142. Online at http://www.newtonproject.sussex.ac.uk/texts/ viewtext.php?id=oth00001&mode=normalized.

to master *all* the apparent confusion: Isaac Newton, Personal Notebook, Pierpont Morgan Library, sheets 5v, 7v, 13r, 15r, 18r, 20v, 28v, 32r–52v. Newton copied much of the material in this notebook from a popular work, *The Mysteries of Nature and Art* by John Bate, published in London in 1654. Newton's use of Bate's book was identified by E. N. da C. Andrade in "Newton's Early Notebook," *Nature* 135 (1935), p. 360, and the connection between Bate and Newton is described in Richard Westfall, *Never at Rest*, p. 61.

6 "I know not what to do": Latin Exercise Book of Isaac Newton, private collection, quoted in Frank Manuel, *A Portrait of Isaac Newton*, pp. 57–58.

"forget his dinner": John Conduitt, Keynes Ms. 130.3, 21r.

his neighbors' grain: William Stukeley, *Stukeley's memoir*, Keynes Ms. 136.03, sheet 6. The stories of Newton's livestock escaping come from manor court records showing fines mulcted on Newton for the offenses. The documents were turned up by Richard Westfall and are quoted in *Never at Rest*, p. 63.

7 more than a mile from Cambridge: William Stukeley, *Stukeley's memoir*, Keynes Ms. 136.03, sheet 7.

"ink to fille it": Isaac Newton, Trinity Notebook, Cambridge Ms. R. 4. 48, f. 1.

milk and cheese, butter and beer: Trinity Notebook, sheets ii–iv. Newton actually listed beer under "Otiose & frustra expensa"—that is, luxuries for which he felt a measure of guilt for indulging in. But as Richard Westfall has also pointed out, many of his contemporaries would have seen beer as essential, or, in Newton's terminology, one of his "Expensa propria."

a man of no social consequence: The question of how poor Newton really was, how subservient he had to be, and how alienated he became as a result is a matter of dispute among leading Newtonians. Richard Westfall argues that Newton's deprivation was real, and the slight truly felt—and Westfall has been echoed by a number of other writers. Mordechai Feingold, professor of the history of science at Caltech, curator of the New York Public Library exhibit *The Newtonian Moment*, and the author of its companion book, challenges that view. Feingold notes, correctly, that Newton was not the recluse he is sometimes painted, and further argues that the sizar status was merely nominal: his allowance was sufficient for such luxuries as the cherries mentioned above, and Newton's family connection to one of Trinity's senior members, a relationship that could have cushioned the worst of his servitor status.

My own view is based on the only three existing sources of anecdotes about Newton's undergraduate years, all written decades after the fact. Beyond those, there is one notebook that contains a partial record of Newton's accounts, and another with the astonishing confession of Newton's sins from 1662 and before. Trinity College records provide the institutional setting for the more personal recollections—and that's all there is. That thin documentary foundation offers great latitude for interpretation. In the end, the question turns on one's own judgments—guesses, really—about human nature in general and Newton's character in particular. Ultimately, as the text above states, I fall more toward the Westfall end of the spectrum: I think the record better supports a picture of a largely solitary young man with no strong emotional or social connection to his classmates, and with some real grounds for resentment and/or envy. But Feingold and others in the current generation of Newton scholars are clearly correct to point out that Newton was not totally friendless, not incapable of ordinary human contact, not averse to all pleasures, including such overtly sensual ones as good food and, on occasion, beer.

just one letter to a college contemporary: Isaac Newton to Francis Aston, 18 May 1699, *Correspondence 1*, document 4, p. 9.

8 not one of the students from his year: Westfall, *Never at Rest*, p. 75. The letter was to Francis Aston, a fellow of Trinity College and later a member and then the secretary (with Robert Hooke) of the Royal Society. See *Correspondence 1*, document 4, pp. 9–11.

"money learning pleasure more than Thee": Fitzwilliam Notebook, sheets 3r–4v.

9 "any demonstrations of them": Abraham de Moivre, "Memorandum relating to Sr Isaac Newton given me by Mr Demoivre," Cambridge Add. Ms. 4007, pp. 706r–707r, cited in D. T. Whiteside, "Sources and Strengths of Newton's Early Mathematic Thought," in Robert Palter, ed., *The Annus Mirabilis of Sir Isaac Newton, 1666–1696*, p. 72.

the meals he forgot to eat: Nicholas Wickens reported on Newton's indifference to sleep and food in a letter to Robert Smith, 16 January 1728, in which he described his father's memories of his chamber mate. Keynes Ms. 137, sheet 2. John Conduitt told the story of the cat's avoirdupois in his memoirs of Newton, Keynes Ms. 130.6, cited in Richard Westfall, *Never at Rest*, pp. 103–4.

10 "except God": Isaac Newton, *Quæstiones quædam Philosophicæ*, Cambridge Add. Ms. 3996, f. 1/88. The quoted material comes from page 338 in the excellent transcript of the *Quæstiones* in J. E. McGuire and Martin Tamny, *Certain Philosophical Questions*, pp. 330–489.

11 the point of his needle: Isaac Newton, Cambridge Add. Ms. 3975, reproduced in Richard Westfall, *Never at Rest*, p. 95.

12 "Great fears of the Sicknesse": Samuel Pepys, *The Shorter Pepys*, p. 486. "which took away the apprehension": Ibid., p. 494.

13 "such a calamity as this": Daniel Defoe, *A Journal of the Plague Year*, pp. 62–63.
"on occasion of the Pestilence": Cited in Richard Westfall, *Never at Rest*, p. 142.

2. "THE PRIME OF MY AGE"

14 "a perfect cure of the plague": Samuel Pepys, *The Shorter Pepys*, p. 557.
"Mathematicks & Philosophy": Cambridge Add. Ms. 3968.41, f. 85, cited in Richard Westfall, *Never at Rest*, p. 143.
infinitesimally small forms: D. T. Whiteside, ed., *The Mathematical Papers of Isaac Newton*, vol. 1, p. 280.
Newton returned to Trinity College: Richard Westfall, *Never at Rest*, p. 142.

15 "rays of gravity": Isaac Newton, *Quæstiones quædam Philosophicæ*, Cambridge Add. Ms. 3996, f. 19.
heavy with fruit: John Conduitt reported a conversation he had with Newton in the last year of his life, 1726. Abraham de Moivre recorded another mention of the apple tree in a memorandum composed in 1727.

16 "thro' the universe": William Stukeley, *Memoirs of Sir Isaac Newton's Life*, pp. 19–21.
A sliver of the old tree: D. McKie and G. R. de Beer, "Newton's Apple: An Addendum," *Notes and Records of the Royal Society of London* 9, no. 2 (May 1952), pp. 334–35.
Newton's apple itself is no fairy tale: D. McKie and G. R. de Beer, "Newton's Apple," *Notes and Records of the Royal Society of London* 9, no. 1 (October 1951), pp. 53–54. Cuttings from Newton's tree have now been propagated in several locations and are for sale by Deacon's Nursery on the Isle of Wight. One cutting, given to my home institution, the Massachusetts

Institute of Technology, produced its first fruit in September 2006, in a small garden next to Building 11. At Woolsthorpe, an old Flower of Kent–bearing tree still drops fruit in the garden there. This tree shoots up from a bent-over section of trunk, presumably the blown-down remnants of the Newtonian original.

17 did not publish his result until 1673: For a more detailed account of Newton's derivation of the formula for centrifugal force, see D. T. Whiteside, "The Prehistory of the *Principia*," *Notes and Records of the Royal Society of London* 45, no. 1 (January 1991), p. 13. Richard Westfall presents a less technical version in *Never at Rest*, pp. 148–50, which draws on J. W. Herivel's article "Newton's Discovery of the Law of Centrifugal Force," *Isis* 51 (1960), pp. 546–53, and his book *The Background to Newton's Principia*.

the centrifugal push: Isaac Newton, *Correspondence 3*, 46–54. See D. T. Whiteside, "The Prehistory of the *Principia, Notes and Records of the Royal Society of London* 45, no. 1 (January 1991), pp. 14–15, for a discussion of Newton's analysis of the motion of a pendulum.

18 would come to be called gravity: This is not quite accurate. Objects under the influence of gravity travel around the center of mass of the total system, not just of the more massive body, as Newton in fact understood. At this time Newton did not have a clear understanding of the concept of inertia, nor yet his first law of motion, which holds that objects at rest or in linear motion tend to stay in motion or at rest unless acted upon by an exterior force. Without this fundamental idea, first suggested to Newton by Robert Hooke in a letter in 1679, his conception of gravity remained imprecise. See Richard Westfall, *Never at Rest*, pp. 382–88, for a discussion of both the insight and the conflict between Hooke and Newton that followed. See also S. Chandrasekhar's summary of the sequence of development of Newton's thinking in the first chapter of his *Newton's Principia for the Common Reader*, pp. 1–14. (Note, however, that I. Bernard Cohen, one of the great Newton scholars of the twentieth century and the translator of the best available English version of the *Principia*, does not think highly of Chandrasekhar's historical skills, and it is true, as Cohen says, that the "common reader" of Chandrasekhar's title had better know a lot of mathematics to make it through most of the argument. Nonetheless, Chandrasekhar, a Nobel laureate in physics, does offer a good introductory summary of the basic concepts in the first section of his book, and it is worth a look.) Another good account of the development of Newton's thoughts on gravity during this period comes in A. Rupert Hall's highly readable biography *Isaac Newton: Adventurer in Thought*, pp. 58–63.

"sixty measured Miles only": William Whiston, *Memoirs of the Life of Mr. William Whiston by himself*, cited in J. W. Herivel, *The Background to Newton's Principia*, p. 65. Whiston was a Newton protégé and his successor to the Lucasian professorship at Cambridge. The circumference of the earth at the equator is 24,902 miles, or 40,076 kilometers. One degree, or ⅟₃₆₀, of that total is 69.172 miles, or 111.322 kilometers. D. T. Whiteside points out that Whiston may not be a reliable source—even though he was Newton's successor as Lucasian Professor, his notes on Newton's life were composed after his predecessor had died. The earliest surviving calculation in Newton's hand that analyzes the motion of the moon subject to an attractive force decreasing with the square of the distance from the earth comes no earlier than 1669, three years after the plague season. Still, it is clear that Newton's work on the problem began while he was sheltering from the epidemic. See D. T. Whiteside, "The Prehistory of the *Principia*," pp. 18–20. My thanks to Simon Schaffer for advice on picking through the minefield of memory and determined fact that stretches across the history of the miracle years. All errors that remain are, of course, mine, not his.

19 "subjecting motion to number": The phrase "subjecting motion to number" is borrowed from Alexander Koyré's marvelous essay "The Significance of the Newtonian Synthesis," which has been published a number of times but is most accessible in I. Bernard Cohen and Richard Westfall, eds., *Newton*, p. 62. I appropriate the line with an apology: Koyré applied it to Galileo, but it is just as true of Newton.

"mutation in its state": J. W. Herivel, *The Background to Newton's* Principia, pp. 157–58.

20 "a Chair to sit down in": Humphrey Newton, Keynes Ms. 135, quoted in Richard Westfall, *Never at Rest*, p. 406. I have corrected a misspelling in the original document of the name Alchimedes and have rendered the "eureka" in English instead of the original's Greek.

the landlocked Newton sought out data: Simon Schaffer has pointed out Newton's use of a global knowledge network in several papers. See Schaffer's "Golden Means: Assay Instruments and the Geography of Precision in the Guinea Trade" and his as yet unpublished "Newton on the Beach," a lecture delivered at Harvard University on April 4, 2006.

3. "I Have Calculated It"

22 "read to yᵉ Walls": Humphrey Newton to John Conduitt (Newton's nephew-in-law and successor as Master of the Mint), 17 January 1727/8, Keynes Ms. 135, f. 2, available at the Newton online archive, http://www .newtonproject.sussex.ac.uk. Newton did not in fact fully satisfy even the minimum requirements of the post. He did lecture regularly on algebra but failed to place any manuscripts in the university library until 1684, when he handed over a text covering eleven years' worth of lectures. The work was later published as *Arithmetica universalis*.

23 "not enduring to see a weed in it": Humphrey Newton to John Conduitt, 17 January 1727/8, Keynes Ms. 135, ff. 1–3.

"failure to reach fulfillment": Richard Westfall, *Never at Rest*, p. 407.

24 "heavenly motions upon philosophical principles": Isaac Newton to Edmond Halley, 27 May 1686, *Correspondence 2*, p. 433.

a book worth forty shillings: In fact, this was a substantial offer, roughly a week's stipend for someone holding an endowed chair at Cambridge. Books were still scarce and valuable enough that the Royal Society would ultimately try to pay Halley for the trouble and expense of his seeing the *Principia* through publication with fifty copies of another Royal Society title, *The History of Fishes*. (The society also tried to do the same to cover Robert Hooke's back wages, but Hooke refused, preferring to wait for the cash he eventually received.)

"as good as his word": Edmond Halley to Isaac Newton, 29 June 1686, *Correspondence 2*, pp. 441–43.

"renew it & send it to him": Newton's account as reported by Abraham de Moivre, a mathematician and Huguenot refugee to London who knew both Halley and Newton. The original is in the Joseph Halle Schaffner Collection of the University of Chicago Library, Ms. 1075–77. The passage is quoted in Alan Cook, *Edmond Halley*, p. 149. There is a slightly fishy smell to Newton's claim to have lost the earlier demonstration, since historians have identified in his papers a document that is either the "lost" original or a version with corrections, which implies that Newton still possessed his earlier calculation when he spoke with Halley. It may be that Newton recognized a flaw in the earlier work, one that he had to correct before showing it to Halley—and certainly before he allowed the dangerous Hooke to see it. Hooke had previously caught a mistake in another demonstration, and Newton had no desire to endure that kind of humiliation again. See Richard Westfall, *Never at Rest*, p. 403.

25 an inverse square attraction: For a marvelous and highly readable account of Newton's proof that inverse square attraction generates Keplerian orbits, see David L. Goodstein and Judith R. Goodstein, *Feynman's Lost Lecture*. In that book, the Goodsteins provide a history of the problem of the shapes of orbits, and then describe how Richard Feynman reconstructed Newton's proof (in its final form, as published in the *Principia*).

only closed path: Formally, a truly circular orbit is possible, as a circle is simply the limit of an ellipse when its eccentricity goes to zero. In physically realistic situations, though, this solution is vanishingly unlikely. Thanks to Caltech's Sean Carroll for pointing out this subtlety.

27 he did not choose to share: See *Correspondence 2*, documents 272, 274, 276, 278, 280, 281, 284.

"all my results together": Isaac Newton, *Principia*, p. 383.

"any motions whatever": Ibid., p. 382.

"to change its state by forces impressed": Ibid., p. 416.

"A change in motion is proportional": Ibid.

28 "always equal and always opposite": Ibid., p. 417.

seemingly simple point of origin: The first edition of the *Principia* contained 510 pages; 210 in Book One, 165 in Book Two, and 110 in Book Three, with the rest given over to introductory remarks and other apparatus.

29 "the Hour of Prayer": Humphrey Newton to John Conduitt, 17 January 1727/8, Keynes Ms. 135, ff. 2–3.

"less to clarify the celestial motions": Isaac Newton, *Principia*, p. 790.

"to demonstrate the other phenomena": Ibid., p. 382.

"from these same principles": Ibid., p. 793.

30 properties of bodies anywhere: In Newton's words, this "rule for the study of natural philosophy" reads: "Those qualities of bodies that cannot be intended and remitted [i.e., qualities that cannot be increased and diminished] and that belong to all bodies on which experiment can be made should be taken as qualities of all bodies universally." Ibid., p. 795 (interpolation by the translators).

31 good data from bad: Simon Schaffer, "Newton on the Beach," pp. 14–17. This unpublished lecture, delivered at Harvard University on April 4, 2006, presents a valuable analysis of Newton's approach to measurement and the significance of the worldwide system of observation and knowledge to Newton's science.

the farthest extremes of the heavens: Isaac Newton, *Principia*, pp. 901–16.

"The theory that corresponds exactly": Ibid., p. 916.

4. "THE INCOMPARABLE MR. NEWTON"

34 a man whom Newton admired: John Locke, *Philosophical Transactions of the Royal Society* 24, no. 298, pp. 1917–20. I am grateful to Jan Golinski for leading me to Locke's weather diary.

confinement to ye London air: Newton to Locke (draft), *Correspondence 3*, p. 184.

35 "No closer to the gods": Isaac Newton, *Principia*, p. 380.

"those that shall succeed him": Edmond Halley, "Accounts of Books," *Philosophical Transactions of the Royal Society* 16, pp. 283–397.

36 "demonstrations more precise": Allen Gabbey, "The *Principia*, a Treatise on 'Mechanics'?" in P. M. Harman and Alan E. Shapiro, eds., *The Investigation of Difficult Things*, p. 306.

"few will understand it": David Gregory to Isaac Newton, 2 September 1687, *Correspondence 2*, p. 484.

37 and so Locke read on: This account almost certainly comes from conversations between Locke and Newton. It has been reported in a number
of places, including in the notes John Conduitt took for his memoirs of
Newton. See, for example, Keynes Ms. 130.5, sheet IV, available at http://
www.newtonproject.imperial.ac.uk/texts/viewtext.php?id=THEM00168&
mode=normalized. For other sources, see Richard Westfall, *Never at Rest*,
pp. 470–71 and footnotes.

"this stupendous Machine": John Locke, "On Education," *Works of John
Locke*, vol. 3, p. 89.

he was a political exile: A warrant for Locke's arrest had been issued in
1685, which prompted the English government to seek his extradition
from Holland. The threat prompted Locke to go at least partly underground in Amsterdam. See Maurice Cranston, *John Locke*, pp. 252–54.

38 "mixture of Papist & Protestants": The phrase appears in a draft paragraph
for a document to be submitted to the Ecclesiastical Commission hearing
the university's claim that it need not obey James's command. The paragraph is quoted in full in Richard Westfall, *Never at Rest*, pp. 478–79.

"Go your way, and sin no more": Ibid., p. 479; the entire incident is covered
on pp. 473–79.

39 his only documented statement: See Ibid., p. 483, and A. Rupert Hall, *Isaac
Newton*, p. 231.

40 a lewd joke about a nun: John Conduitt, Keynes Ms. 130.6, Book 2, cited
in Richard Westfall, *Never at Rest*, p. 192.

In contrast, Locke played politics. Maurice Cranston, *John Locke*, p. 219.

41 effervescent burst of creation: I was first directed to Hooke's and Locke's
interest in weather measurement by Jan Golinski, and Professor Golinski's
new *British Weather and the Climate of Enlightenment* offers the best account of that pursuit in the context of broader currents in the enlightenment. For a very brief summary of Hooke's weather work, see "A History
of the Ecological Sciences, Part 16: Robert Hooke and the Royal Society
of London," *Bulletin of the Ecological Society of America*, April 2005, p. 97.
But for the bitter conflicts with Newton, Hooke would be remembered
unequivocally as what he was: one of the most extraordinary figures in
the history of science. The breadth of his interests and accomplishments
earned him the nickname "London's Leonardo," and there is a good measure of truth in the epithet.

43 "as welcome as I can make you": Isaac Newton to John Locke, 28 October
1690, *Correspondence 3*, p. 79, and 3 May 1692, *Correspondence 3*, p. 214.

a simplified version of the proof: Isaac Newton to John Locke, March
1689/90, *Correspondence 3*, pp. 71–77.

"his never enough to be admired book": John Locke, *Essay on Human
Understanding*, Book 4, chapter 7, paragraph 11 (3).

44 "beyond the Reach of humane Art & Industry": Humphrey Newton to John Conduitt, 17 January 1727/8 and 14 February 1727/8, Keynes Ms. 135, sheets 3, 5.
"neither he nor any body else understands": John Conduitt reported this story, which is cited in Robert Westfall, *Never at Rest*, p. 486.

45 "thanks to my Lord and Lady Monmoth": Isaac Newton to John Locke, 14 November 1690, *Correspondence 3*, p. 82. Charles Mordaunt, Earl of Monmouth, was one of William's most powerful English supporters.
"my most humble service": Isaac Newton to John Locke, 13 December 1691, *Correspondence 3*, pp. 185–86.

5. "The Greatest Stock of Impudence"

50 a family coining business: The underlying source for almost all of Chaloner's biography before his conflict with Newton is the pamphlet by an anonymous author titled *Guzman Redivivus: A Short View of the Life of Will. Chaloner*, 1699. Additional details come from the *Oxford Dictionary of National Biography* entry "Chaloner, William," by Paul Hopkins and Stuart Handley.
"some unlucky Rogues Trick or other": *Guzman Redivivus*, p. 1.
an agreed number of nails: See H. R. Schubert, *History of the British Iron and Steel Industry from c. 450 B.C. to A.D. 1775*, pp. 304–12. There is a diagram of a water-powered slitting mill on p. 309.

51 "St. Francis's Mule": *Guzman Redivivus*, p. 1. The phrase "St. Francis's Mule" is a play on the story in the saint's life in which, on seeing their master becoming increasingly infirm, Francis's followers "borrow" (read, "steal") a mule. The animal's owner, on seeing Francis astride his mule, rebukes the saint, telling him to try to be as virtuous as others believe him to be. Francis dismounts from the mule and kneels before its owner, thanking him for the advice before, presumably, continuing on foot.
"with a purpose to visit London": *Guzman Redivivus*, p. 1.

52 Its death rate exceeded the birth rate: See, for example, the bill of mortality for London for the year 1700, reproduced in facsimile on page 97 of Maureen Waller, *1700: Scenes from London Life*. In that year, the company of parish clerks for London reported 14,639 christenings and 19,443 deaths. For data on London's population history in the seventeenth century, see R. A. Houston, "The Population History of Britain and Ireland, 1500–1750," in Michael Anderson, ed., *British Population History*, pp. 118–24, and David Coleman and John Salt, *The British Population*, pp. 27–32. For more details on the social structure of the migration to London and comparisons to other cities, see Roy Porter, *London: A Social History*, pp. 131–33, and Stephen Inwood, *A History of London*, pp. 269–75.

"region of dirt, stink and noise": Arthur Young, cited in Roy Porter, *London: A Social History*, p. 133.

heaps of human and animal waste: London had some open sewers, but not closed ones at this time. The Houses of the middle class and above had vaults for their waste, which would be cleaned out by nightsoil men, though the system was far from foolproof. Serious attempts at sanitation did not come to London until after 1750. Probably not coincidentally, birth rates began to exceed death rates in the metropolis after that date. See Frank McLynn, *Crime and Punishment in Eighteenth-Century England*, p. 2.

"come tumbling down Flood": Jonathan Swift, "A Description of a City Shower," *The Tatler*, October 1710, quoted in Stephen Inwood, *A History of London*, p. 282.

53 "impure and thick mist": Jonathan Evelyn in *Fumifugium*, published in 1661, cited in Roy Porter, *London: A Social History*, p. 97. See also Maureen Waller's *1700: Scenes from London Life*, pp. 95–96.

wages in the capital: Roy Porter, *London: A Social History*, p. 132.

54 exhilarating, terrifying, incomprehensible: Ibid., p. 134.

the proper coffeehouse: Jan Golinksi, personal communication.

most sophisticated houses of prostitution: Frank McLynn, *Crime and Punishment in Eighteenth-Century England*, p. 99.

"at a loss of Acquaintance": *Guzman Redivivus*, p. 1.

London hatters beat to death a man: Stephen Inwood, *A History of London*, p. 334. Inwood drew this anecdote from J. Rule, *The Experience of Labor*, p. 111.

55 whatever price the cartel set: Stephen Inwood, *A History of London*, p. 325.

broke, bored, or both: For an account of the dashing career of Thomas Butler, wealthy spendthrift, Jacobite spy, and eminently respectable highwayman, along with an account of the social and mythic status of highwaymen, see Frank McLynn, *Crime and Punishment in Eighteenth-Century England*, pp. 55–60.

56 Others preyed on coaches: Ibid., pp. 5–6.

"yet being so warily watchful": Robert Greene, "A Disputation Between a He Cony Catcher and a She Cony Catcher," pp. 211–12, cited in John L. McMullan, *The Canting Crew*, p. 101.

Fences, at the hub of criminal gangs: Ibid., pp. 105–7.

57 "looser associates": *Guzman Redivivus*, p. 1.

"a little of your plain fucking": Arbuthnot's quip was recorded in Horace Walpole's correspondence, vol. 18, p. 70, of the edition edited by W. S. Lewis, cited in Frank McLynn, *Crime and Punishment in Eighteenth-Century England*, p. 99.

58 a kind of affordable fashion accessory: My thanks to the horological expert Will Andrewes (formerly curator of Harvard University's Museum of Historical Scientific Instruments), the British Museum's David Thompson, and the Museum of London's Hazel Forsyth for their help in narrowing down Chaloner's likely invention.

59 jail fever: Maureen Waller, *1700: Scenes from London Life*, pp. 101–2.
    lost about two-thirds of their children: For the figures on Quaker mortality, see J. Landers, "Mortality and Metropolis: The Case of London, 1675–1875," *Population Studies* 41, no. 1 (March 1987), p. 74. A good summary of the general background of disease and mortality in the city can be found in Stephen Inwood, *A History of London*. See also J. Landers, "Burial Seasonality and Causes of Death in London, 1670–1819," *Population Studies* 42, no. 1 (March 1988), pp. 59–83.
    "the Character of his Servant": *Guzman Redivivus*, p. 2.

60 "discovering Stol'n Goods &c": Ibid.
    Wild managed this balancing act: For a good summary of Wild's career, see Frank McLynn, *Crime and Punishment in Eighteenth-Century England*, pp. 22–30. For a marvelous fictional imagining of Wild's world, see David Liss's well-researched novel *A Conspiracy of Paper*.
    "some Old Garret to repose his Carcase": *Guzman Redivivus*, p. 2.

6. "Every Thing Seem'd to Favour His Undertakings"

61 "daubed with colours": Isaac Newton, "Chaloner's Case," Mint 19, I, f. 501.

62 "As o'er-dy'd blacks": Shakespeare, *The Winter's Tale*, act 1, scene 2.

63 a mere 103: Lord Macaulay, *The History of England*, vol. 5, p. 2566.
    "the spectacle of his death": Ibid., p. 2564. Macaulay commends the extraordinary diary of Narcissus Luttrell as a source to give a sense of the ubiquity of clipping and of crime in general—and so do I.

64 "88.8 grains of silver": This number comes from Newton's report to the Treasury, 25 September 1717, on the metal content of English money. He states: "I humbly represent that a pound weight Troy of Gold, eleven ounces fine & one ounce allay, is cut into 44½ Guineas, & a pound weight of silver, 11 ounces, 2 pennyweight fine, & eighteen pennyweight allay is cut into 62 shillings, & according to this rate, a pound weight of fine gold is worth fifteen pounds weight six ounces seventeen pennyweight & five grains of fine silver, recconing a Guinea at 1£, 1s. 6d. in silver money."

65 "freer from clipping or counterfeiting": Samuel Pepys, *Diary*, vol. 10, online at http://www.gutenberg.org/etext/4127. Further details in this description come from Sir John Craig, *Newton at the Mint*, pp. 5–8, and from C. E. Challis, *A New History of the Royal Mint*, pp. 339–48.
    The 1662 transition to a milled coinage using Blondeau's milling machinery was not the first time the Mint produced milled coins. Milled

coins were made under Queen Elizabeth I for ten years, using horse-driven mills in a process supervised by another Frenchman, Eloye Mestrell. The coins, both gold and silver, were of far higher quality than the hammered coinage, but the procedure was slow, and after ten years Mestrell was dismissed and the use of machinery to produce counterfeit-resistant coins was dropped. (Mestrell was later hanged for counterfeiting.) Next, beginning in 1631, Nicholas Briot, yet one more French currency engineer, and then his son-in-law John Falconer, supervised the making of several runs of milled coins in gold and silver for the English and the Scottish mints. These coinages also failed to supersede the traditional hammered currency, a state of affairs that would continue until Blondeau managed to prove the effectiveness of his machinery, beginning with his demonstration to Cromwell in 1656. See Challis, pp. 250–51 for Mestrell's story, and pp. 300–302 and 339 for the Briot-Falconer details.

66 "Coining any manner of Money": "Inhabitant, offences against the king: coining," 1 September 1686, *Proceedings of the Old Bailey,* http://www .oldbaileyonline.org.

a better choice in her confederates: "Mary Corbet, offences against the king: coining," 9 April 1684, *Proceedings of the Old Bailey,* http://www .oldbaileyonline.org.

67 A jury of women was impaneled: Summary of sessions on 9 April 1684, *Proceedings of the Old Bailey,* http://www.oldbaileyonline.org. "Pleading the belly" was a traditional and much-used tactic to delay execution. The standard for pregnancy was whether a jury of women ("matrons") could detect a "quickened" child—i.e., a fetus with detectable movements; if so, the execution would be delayed until delivery. (This is how Moll Flanders describes her beginnings in Daniel Defoe's eponymous novel.) In many cases, the reprieve led to a pardon or commutation of sentence. Some obvious consequences resulted from this provision of the common law: John Gay's *The Beggar's Opera,* first produced in 1765, includes a character who worked as a "child-getter," who kindly offered his services to ladies in prison in fear of their lives. The right to plead a pregnancy was formally abolished in Great Britain only in 1931.

68 "as guineas go now": "Samuel Quested, Mary Quested, J—C—, offences against the king: coining," *Proceedings of the Old Bailey,* 14 October 1695, http://www.oldbaileyonline.org. At trial, Samuel Quested took all the responsibility on himself, and Mary won an acquittal on his testimony despite claims of witnesses that she was involved in the coining and clipping side of the family business. Samuel had no chance to survive his arrest. His long record as a coiner and the dangerous excellence of his work doomed him. The trial jury convicted him of high treason, and he was hanged.

69 Chaloner found his die maker: Isaac Newton, "Chaloner's Case," Mint 19/1. See also "The Information of Katherine Carter the wife of Thomas

Carter now prisoner in Newgate 21th. February 1698/9," Mint 17, document 120—one of the sources on which Newton based his account.

solar eclipse: The eclipse broadside is extremely rare. The British Museum has one copy in its collection. Taylor's collections of maps are more common, with individual sheets appearing for sale from time to time by antiquarian dealers.

70 a thin skim of gold: "The Information of Math[ew] Peck of Pump Court in Black Fryars Turner 25 day of January 1698/9," Mint 17, document 117. See also "The Deposition of Humphrey Hanwell of Lambeth p[ar]ish in Southwark 22d. Feb[ruar]y 1698/9," Mint 17, document 123.

"boast his workmanship": Isaac Newton, "Chaloner's Case," Mint 19/1, f. 501.

"daily falling into his lap": *Guzman Redivivus*, p. 2.

71 "she went to bed with her Spark": Ibid., p. 3.

"Guinea's and Pistoles": Ibid., p. 3.

he stashed his precious dies: Isaac Newton, "Chaloner's Case," Mint 19/1, f. 501.

73 Butler and Newbold: The quotes are from *Guzman Redivivus*, p. 4, except for the claim that Chaloner used threats and money to sway the printers, which comes from Isaac Newton's account of the same incident in "Chaloner's Case," Mint 19/1, f. 501. It is not clear whether Chaloner's dupes were ever executed. The author of *Guzman* reported that the two convicted traitors remained in prison at the time of Chaloner's own execution in 1699. This seems unlikely, implying as it does a six-year stint in one of the most harsh and disease-ridden environments in London—but it is possible. Newton did not report on the fate of the printers.

"He had fun'd": Isaac Newton, "Chaloner's Case," Mint 19/1, f. 501.

his cash flow ebbed: Paul Hopkins and Stuart Handley, "Chaloner, William," *Oxford Dictionary of National Biography*.

74 He was pronounced guilty: "Matthew Coppinger, theft : specified place, theft with violence : robbery, miscellaneous : perverting justice," 20 February 1695, *Proceedings of the Old Bailey*, http://www.oldbaileyonline.org.

"Relations, nay": *Guzman Redivivus*, p. 5.

7. "All Species of Metals . . . from This Single Root"

77 sickly as a child: Robert Boyle was born on 25 January 1626/7.

78 the leader of the nation's intellectual life: The record of Newton's movements and the quote from Samuel Pepys's letter to John Evelyn on 9 January 1692 both come from Richard Westfall, *Never at Rest*, p. 498.

"Mr. Boyles red earth": Isaac Newton to John Locke, 16 February 1691/2, *Correspondence 3*, p. 195.

79 the red earth was ultrasensitive: Isaac Newton to John Locke, 7 July 1692, *Correspondence 3*, p. 215.
"may lie concealed": John Locke to Isaac Newton, 26 July 1692, *Correspondence 3*, p. 216.

80 "as good Dutch dollars": Ben Jonson, *The Alchemist*, act 3, scene 2, http://www.gutenberg.org, 10th ed., May 2003.

81 "except God": Isaac Newton, *Quæstiones quædam Philosophicæ*, Cambridge Add. Ms. 3996, f. 1/88. The quoted material comes from page 338 in the transcription of the *Quæstiones* in J. E. McGuire and Martin Tamny, *Certain Philosophical Questions*, pp. 330–489.

82 "the serpent that hides": Martin Schoock, *Admironda Methodus Novae Philosophiae Renati Descartes* (1643), p. 13, quoted in Desmond Clarke, *Descartes*, p. 235. See also Michael Heyd, *"Be Sober and Reasonable,"* pp. 123–24.
"I secretly teach atheism": René Descartes to M. de la Thuillerie, 22 January 1644, quoted in Desmond Clarke, *Descartes*, p. 240.
"may the Sun imbibe this Spirit": Isaac Newton to Henry Oldenburg, 7 December 1675, *Correspondence 1*, document 146, p. 366.

83 "the life of all things": Isaac Newton, *Principia*, translated by I. B. Cohen and Anne Whitman, p. 926. I am grateful to Professor Schaffer for pointing out to me this and the previous reference.
"useful for that purpose": Isaac Newton to Richard Bentley, 10 December 1692, *Correspondence 2*, document 398, p. 233.
"all ear, all brain, all arm": Isaac Newton, *Principia*, p. 942.
"Parts of the Universe": Isaac Newton, *Opticks*, p. 403.

84 "her fire, her soule, her life": "The Vegetation of Metals," Burney Ms. 16, f. 5v. I am indebted to Richard Westfall's description of Newton's early alchemical phase for much of my account here. See his *Never at Rest*, pp. 304–9.
changes of the living world: I draw this interpretation of Newton's alchemy primarily from the writing of Betty Jo Teeter Dobbs. See especially *The Janus Faces of Genius*, and in that work, chapter 4, "Modes of Divine Activity in the World: Before the *Principia*," pp. 89–121.

85 "the illumination of matter": Isaac Newton quoted in Jan Golinski, "The Secret Life of an Alchemist," in the invaluable collection *Let Newton Be!*, edited by John Fauvel et al., p. 160.

8. "THUS YOU MAY MULTIPLY TO INFINITY"

86 his alchemical laboratory: The image is one of those made by David Loggan and published in his *Cantabridgia illustrata* in 1690. It can be seen at http://upload.wikimedia.org/wikipedia/commons/thumb/8/87/Trinity_College_Cambridge_1690.jpg/800px-Trinity_College_Cambridge_1690.jpg.

breach of security: "Of the Incanlescense of *Quicksilver with Gold,* generously imparted by B. R. [Robert Boyle], *Philosophical Transactions of the Royal Society* 10 (1676), 515–33.

Act Against Multipliers: The law was enacted in 1404, during the reign of King Henry IV, and it stated "that none from henceforth should use to multiply gold or silver, or use the craft of multiplication; and if any the same do, they incur the pain of felony." Newton refers to Boyle's role in the repeal of the act in a letter to John Locke, 2 August 1691, *Correspondence 2,* p. 217.

87 "will sway him to high silence": Isaac Newton to Henry Oldenburg, 26 April 1676, *Correspondence 2,* p. 2.

alchemical reactions: For an example of his involvement in the making of his apparatus, see the drawing in Newton's hand of several furnaces, reproduced in Richard Westfall, *Never at Rest,* p. 283; the catalogue of his books and its implications are discussed in John Harrison's *The Library of Isaac Newton;* Newton's alchemical notes total a formidable record. Most of the major texts are housed in the Keynes Collection at King's College, Cambridge, but the traditional starting place to gain an acquaintance with Newton's alchemical procedure is the Cambridge University Library's Add. Ms. 3975, which has been transcribed with reproductions of Newton's diagrams and is available at http://webapp1.dlib.indiana.edu/newton/mss/norm/ALCH00110. The site contains, among much else, both the records of experiments and examples of his reading notes from other alchemical authorities. Add. Ms. 3973 also records a long series of alchemical experiments. Humphrey Newton reported on Newton's involvement in experimental practice, and Newton's notes in Add. Ms. 3975 and elsewhere copiously document that commitment.

to the limit of his instruments: Here again, Cambridge Add. Ms. 3975 provides the easiest route to an appreciation for Newton's astonishing capacity for pure, hard labor and his commitment to rigor in his alchemical research as in all his other investigations.

"not able to penetrate": Humphrey Newton to John Conduitt, 17 January 1727/8, Keynes Ms. 135, pp. 2–3.

the true course of his work: Richard Westfall points out this basic fact in *Never at Rest,* pp. 359–61. As Westfall documents, if you look at the time spent on alchemy versus time spent on physics during the early and middle eighties, alchemy wins. Results were a different matter, and Newton was certainly aware that the *Principia* represented a different order of outcomes than anything he had achieved in the laboratory to date.

88 experimental notes: Betty Jo Teeter Dobbs, *The Janus Faces of Genius,* p. 171. *Index Chemicus: Index Chemicus,* Keynes Ms. 30a, can be found at http://webapp1.dlib.indiana.edu/newton/mss/dipl/ALCH00200/. The document is discussed briefly in Richard Westfall's *Never at Rest,* pp. 525–26,

and is analyzed in much greater depth in Westfall's "Isaac Newton's Index Chemicus," *Ambix* 22 (1975), pp. 174–85. The reference to Boyle comes on f. 16r of the document and concerns the specific interaction of gold and "living water" (*aqua animam*).

89   "it amagalms very easily": Isaac Newton, "Experiments and Observations Dec. 1692–Jan. 1692/3," Portsmouth Collection, Cambridge Add. Ms. 3973.8, transcribed in Betty Jo Teeter Dobbs, *The Janus Faces of Genius*, pp. 290–91.

"infernal secret fire": Isaac Newton, *Praxis*, Babson Ms. 420, Sir Isaac Newton Collection, Babson College Archives, transcribed and translated by Betty Jo Teeter Dobbs, *The Janus Faces of Genius*, pp. 293–305. The passages quoted occur on pp. 299–300.

wingless dragons: The history of Newton's papers is summarized neatly at the Newton Project website, http://www.newtonproject.ic.ac.uk. As discussed there, the rehabilitation of Newton's alchemy began when the economist John Maynard Keynes bought most of the available alchemical papers in the famous (in Newton circles) Sotheby's sale of 1936. Keynes wrote the first important essay on Newton's alchemy, thus breaking the taboo that guarded the image of the foundational scientific genius for two centuries. Keynes did not get the *Praxis* manuscript, however. Charles Babson, an American Wall Street magnate with a Newton fascination, outbid him—which is why the manuscript is in the library of Babson College in Massachusetts.

91   "Thus you may multiply to infinity": Isaac Newton, *Praxis*, transcribed and translated by Betty Jo Teeter Dobbs, *The Janus Faces of Genius*, pp. 301–2, 304. The earlier draft cited is quoted in Richard Westfall, *Never at Rest*, p. 530. In the final version, as Dobbs translates it, Newton backs away from his claim of infinite multiplication, writing only that "you may increase it much more" (p. 304).

"quintessential matter": Isaac Newton, *Praxis*, transcribed and translated by Betty Jo Teeter Dobbs, *The Janus Faces of Genius*, p. 305.

9. "Sleeping Too Often by My Fire"

93   and then simply stopped: Isaac Newton, *Praxis*, transcribed and translated by Betty Jo Teeter Dobbs, *The Janus Faces of Genius*, pp. 304–5.

The draft ends in midsentence: Isaac Newton to Otto Mencke (draft), 30 May 1693, *Correspondence 3*, pp. 270–71. Newton did complete and send a letter to Mencke on 22 November 1693, *Correspondence 3*, pp. 291–93.

"distemper": Isaac Newton to John Locke, 15 October 1693, *Correspondence 3*, p. 284, quoted in greater detail below.

94   "my former consistency of mind": Isaac Newton to Samuel Pepys, 13 September 1693, *Correspondence 3*, p. 279.

"better if you were dead": Isaac Newton to John Locke, 16 September 1693, *Correspondence 3*, p. 280.

95 "I remember not": Isaac Newton to John Locke, 15 October 1693, *Correspondence 3*, p. 284. Newton says that he does not remember what he said of Locke's book in the previous letter; he had written that he begged Locke's pardon "for representing that you struck at ye root of morality in a principle you laid down in your book of ideas ... and that I took you for a Hobbist." It is a pleasing detail that Newton, and presumably Locke, saw the charge of Hobbism as at least as serious as that of pandering.

Huygens told Leibniz: See Richard Westfall, *Never at Rest*, p. 535. Westfall draws his account of this gossip-mongering from a letter Huygens wrote to his brother and from an entry in his journal.

"very ill circumstances": John Wallis to Richard Waller, 31 May 1695, *Correspondence 4*, p. 131.

96 "no mortal may approach": Edmond Halley, "Ode to Newton," published as a dedicatory preface to the first edition of the *Principia*. The Latin original of the line reads, "*Nec fas est propius mortali attingere divos.*"

"I was frozen stiff": The error Fatio caught—a mistake in Book Two of the *Principia* on the behavior of water flowing out of a hole in the bottom of a tank—is discussed in the notes to a letter from Fatio to Huygens, *Correspondence 3*, pp. 168–69. Fatio wrote of his plans for a second edition of the *Principia* in a letter to Huygens dated 18 December 1691; the letter and the quote from Fatio's confession of Newton's mathematical superiority are both cited in Richard Westfall, *Never at Rest*, p. 495.

97 "with all my heart": Nicholas Fatio de Duillier to Isaac Newton, 24 February 1689/90, *Correspondence 3*, pp. 390–91.

Newton's secretary: Richard Westfall, *Never at Rest*, p. 496.

Fatio had remained silent: Isaac Newton to John Locke, 28 October 1690, *Correspondence 3*, p. 79.

98 "a heap of trees": Nicholas Fatio de Duillier to Isaac Newton, 10 April 1693, *Correspondence 3*, pp. 265–66.

"the Roman Empire": Nicholas Fatio de Duillier to Isaac Newton, 30 January 1692/3, *Correspondence 3*, pp. 242–43.

99 "what may befall me": Nicholas Fatio de Duillier to Isaac Newton, 17 September 1692, *Correspondence 3*, pp. 229–30.

"with my prayers for your recovery": Isaac Newton to Nicholas Fatio de Duillier, 21 September 1692, *Correspondence 3*, p. 231.

"most humble, most obedient": Nicholas Fatio de Duillier to Isaac Newton, 30 January 1692/3, *Correspondence 3*, pp. 231–33.

100 "you should return hither": Isaac Newton to Nicholas Fatio de Duillier, 24 January 1692/3, *Correspondence 3*, p. 241.

"I am ready to do so": Nicholas Fatio de Duillier to Isaac Newton, 30 January 1692/3, *Correspondence 3*, pp. 242–43.

"wth all fidelity": Isaac Newton to Nicholas Fatio de Duillier, 14 February 1692/3, *Correspondence 3*, p. 244.

"to be free of an excrescence": Nicholas Fatio de Duillier to Isaac Newton, 9 March 1692/3, *Correspondence 3*, pp. 262–63.

"I have not now any need": Nicholas Fatio de Duillier to Isaac Newton, 4 May 1693, *Correspondence 3*, pp. 266–67.

101 "all manner of respect": Nicholas Fatio de Duillier to Isaac Newton, 18 May 1693, *Correspondence 3*, pp. 267–70.

"to bring us out of distresses": Isaac Newton, Fitzwilliam Notebook, pp. 3r–4v, the Newton Project, http://www.newtonproject.sussex.ac.uk.

102 thirty pounds in 1710: Richard Westfall, *Never at Rest*, p. 539.

"I know not what to doe": Latin Exercise Book of Isaac Newton, private collection, quoted in Frank Manuel, *A Portrait of Isaac Newton*, pp. 57–58.

103 Pepys did not respond: Pepys did write to John Millington, to ask if he knew what was wrong with Newton, as he feared "a discomposure in head, or mind, or both." (Pepys to Millington, 26 September 1693, *Correspondence 3*, p. 281.) Millington wrote back that he had visited Newton, who told him that he had written the letter during "a distemper that much seized his head and that kept him awake for five nights together." Millington added that Newton wanted to beg Pepys's pardon for the rudeness—a reply that satisfied Pepys. (Millington to Pepys, 30 September 1693, *Correspondence 3*, pp. 281–82.)

"the same good will": John Locke to Isaac Newton, 5 October 1693, *Correspondence 3*, p. 283.

how Pepys should place his bets: Pepys posed the question in one letter, Samuel Pepys to Isaac Newton, 22 November 1693, *Correspondence 3*, pp. 293–94. Newton replied almost immediately, Isaac Newton to Samuel Pepys, 26 November 1693, *Correspondence 3*, pp. 294–96.

104 problems in calculus: Isaac Newton to Nathaniel Hawes, 25 May 1694, and Newton to Hawes, 26 May 1694, *Correspondence 3*, pp. 357–68; Isaac Newton to E. Buswell, June 1694, *Correspondence 3*, p. 374; and a manuscript from July 1694, *Correspondence 3*, pp. 375–80.

"as the lion is recognized by his print": Richard Westfall, *Never at Rest*, pp. 582–83. Newton's niece, Catherine Barton, reported that it took him just twelve hours to solve the two problems.

10. "THE UNDOING OF THE WHOLE NATION"

110 55,000 pounds sterling: The goldsmiths' figures are reported in Ming-hsun Li, *The Great Recoinage of 1696 to 1699*, p. 53, and the Mint coinage statistic comes from the same source, p. 48. Li draws the latter from Hopton Haynes's records of the annual figures for Mint production in Haynes, *Brief Memoires Relating to the Silver and Gold Coins of England*. The con-

version of weight to number is based on the statutory requirement that the Mint coin three pounds two shillings out of each pound of silver alloy that was about seventy-two-percent-pure silver. The goldsmiths' petition does not specify the fineness or purity of the bullion being exported. It is thus conceivable that the bullion leaving the country contained a higher proportion of silver than minted money did, increasing the loss to England's money supply.

III "who do any thing for their profit": Ming-hsun Li, *The Great Recoinage of 1696 to 1699*, p. 55.

112 "Great masses were melted down": Lord Macaulay, *The History of England*, vol. 5, p. 2564.

A laborer's daily wage: The price and wage numbers come from the data series collected by Gregory Clark as part of the research underpinning several publications. See especially "The Long March of History: Farm Wages, Population and Economic Growth, England, 1209–1869," *Economic History Review*, February 2007, pp. 97–135; Gregory Clark, "The Condition of the Working Class in England, 1209–2004," *Journal of Political Economy* 113, no. 6 (December 2005), pp. 1307–1340. Clark's database, updated most recently on April 10, 2006, can be found at the International Institute of Social History's website, http://iisg.nl/hpw/data.php#united.

The Mint had produced: Hopton Haynes, *Brief Memoires Relating to the Silver and Gold Coins of England*, quoted in Ming-hsun Li, *The Great Recoinage of 1696 to 1699*, p. 48.

113 "Inconveniences to the Kingdom": *Journal of the House of Commons*, May 7, 1690, quoted in Ming-hsun Li, *The Great Recoinage of 1696 to 1699*, pp. 55–56.

"the coins went on dwindling": Lord Macaulay, *The History of England*, vol. 5, p. 2570.

William faced the prospect of being defeated: The conflict has sometimes been called the second Hundred Years' War, and it included campaigns on every continent except Australia and Antarctica. It began in 1688, with what has been called the Nine Years' War, or the War of the Grand Alliance, and continued, with interruptions, through what Europeans call the Seven Years' War and North Americans call the French and Indian War (more properly the fourth French and Indian War, waged from 1749–1756). It went on from there to the American Revolution, in which French intervention was crucial to the defeat of the British, and then the Napoleonic Wars, ending in the British and allied victory at Waterloo in 1815, and which finally settled the question of French territorial ambitions on the European continent.

114 about 100,000 by the mid-1690s: John Childs, *The Nine Years' War and the British Army, 1688–1697*, p. 1.

a professional civil service: For discussions of the politics and implications of the decisions of the Convention Parliament, see as starting points Tony Claydon, *William III*, pp. 60–82, and J. R. Jones, *The Revolution of 1688 in England*, pp. 311–31. For a good account of the artful confusion at the heart of Parliament's grant of the throne to William as well as to his wife, see Howard Nenner's essay "Pretense and Pragmatism: The Response to Uncertainty in the Succession Crisis of 1689," in Lois G. Schwoerer, ed., *The Revolution of 1688–1689*.

appointed tax commissioners: "An Act for Granting an Aid to Their Majesties of the Sum[m]e of Sixteene hundred fiftyone thousand seven-hundred and two pounds eighteen shillings towards the Carrying on a Vigorous Warre against France," *Statutes of the Realm*, vol. 6, http://www.british-history.ac.uk/report.asp?compid=46359. Curiously, there is a William Chaloner listed as a commissioner, but as he was supposed to serve "For the North Rideing of the County of Yorke," it seems unlikely that even so ingenious a scammer of the government as the coining William Chaloner could have wangled an appointment that far from his home haunts in London.

1.2 million pounds: Ming-hsun Li, *The Great Recoinage of 1696 to 1699*, p. 34.

115   exceeded government tax income: D. W. Jones, *War and Economy*, p. 11. In fact, the cost of the Nine Years' War (also known as the War of the Grand Alliance) demonstrated that the military mobilizations both William and Louis had attempted exceeded the capacity of their states to sustain. For France and Great Britain (as the state was called after the Acts of Union between England and Scotland in 1706 and 1707) the armies of the Nine Years' War were the largest either nation fielded until the Napoleonic Wars. It was just too damn expensive.

English silver currency in Amsterdam: Ming-hsun Li, *The Great Recoinage of 1696 to 1699*, p. 58.

English government finances: Richard Hill to Trumbull, 21 August 1695, quoted in John Childs, *The Nine Years' War and the British Army, 1688–1697*, p. 297.

116   "a matter of so general concern": King William to the House of Commons, 26 November 1695, in the *Journal for the House of Commons* vol. 11, p. 339.

11. "Our Beloved Isaac Newton"

118   "to prevent the Melting or Exporting": Isaac Newton, Goldsmiths' Library Ms. 62, quoted in Ming-hsun Li, *The Great Recoinage of 1696 to 1699*, p. 217.

119   "the Party offending": Ibid.

"Change of Denomination": John Locke's response to Lowndes, Gold-smiths' Library Ms. 62, quoted in Ming-hsun Li, *The Great Recoinage of 1696 to 1699*, p. 227.

120 "the measure of the bargain": John Locke quoted in Ming-hsun Li, *The Great Recoinage of 1696 to 1699*, p. 102.

"to defraud the King": Ibid., p. 104.

Newton simply stayed put: In all those years at Cambridge—in all his life, in fact—Newton, the man who solved the motion of the tides (and lived on an island) never once saw the sea. This was pointed out to me by Simon Schaffer, who also included it in the text of a lecture titled "Newton on the Beach," delivered at Harvard University on April 4, 2006.

121 March 21: Richard Westfall, *Never at Rest*, p. 556.

"more attendance than you may spare": *Correspondence 4*, document 545, p. 195.

"the office of Warden of the Mint": *Correspondence 4*, document 547, p. 200.

Not a single letter: Richard Westfall, *Never at Rest*, p. 550.

12. "STIFLING THE EVIDENCE AGAINST HIM"

122 "Double Deception": *Guzman Redivivus*, p. 7.

failing money supply: John Locke's public writings on the currency (as distinct from the argument produced in response to Lowndes's request of 1695) include *Some Considerations of the Consequences of the Lowering of Interest and Raising the Value of Money* (1691), *Short Observations on a Printed Paper entitled 'For Encouraging the Coining Silver Money in England and after keeping it here'* (1695), and *Further Considerations Concerning Raising the Value of Money* (1695). Charles Davenant produced *An Essay on the Ways and Means of supplying the War* (1694).

123 set their thoughts in type: Sir Robert Cotton, *Touching the Alteration of Coin* (London: Thomas Horne, 1690); John Briscoe, *Proposals for Supplying the government with Money on easie Terms . . .* (London, 1694); and J. C. Merchant, *Proposals for regulating the silver coyne, bearing the charge of it, producing a circulation, and securing it to the Kingdom* (1695) can be found in the Goldsmiths' Library, London, now housed at the Senate House Library and reproduced in an online resource: The Making of the Modern World: The Goldsmiths'-Kress Library of Economic Literature, 1450–1850. Joyce Oldham Appleby, in her essay "Locke, Liberalism and the Natural Law of Money," *Past and Present*, no. 71 (May 1976), p. 46, gives another brief survey of tracts on money, including *The Groans of the Poor* and *A letter from an English Merchant at Amsterdam to his Friend at London*.

paper as a tool of thought: I am indebted to James Gleick's observation on the scarcity of paper in Newton's childhood for leading me to this line of thought. See Gleick's *Isaac Newton*, p. 14.

coarse brown paper: D. C. Coleman, *The British Paper Industry*, pp. 41–43.
eighty thousand reams: See the table compiled by D. C. Coleman, *The British Paper Industry*, p. 13. The figures there are incomplete for the early years of the series, sometimes showing the amount imported by English merchants (e.g., 1621 and 1626) and sometimes the total by foreign traders (e.g., 1622 and 1624). I have combined the separate totals in those years to arrive at my estimate of about eighty thousand reams.
the cost of the printing: Ibid., pp. 11–12.

124 a hundred English mills: Ibid., pp. 49, 56.
deficiencies in the coinage: Paul Hopkins and Stuart Handley, "Chaloner, William," in the *Oxford Dictionary of National Biography*.

125 almost certainly no coincidence: *Proposals for a Fund of A Hundred and Fifty Thousand Pounds per Annum*, pp. 6–7; *The True Way of Taxing shewing What is the Legal* Rack-Rent *for Taxing first of Laymen, secondly of Churchmens Real Estates Equally*, p. 1. In *Proposals for a Fund . . .* the author also suggests increasing government revenue by a new system of fines for criminals—which would surely have offended Chaloner's sense of self-interest, had he known of the tract and/or the idea.

127 four in Birmingham and Sheffield: William Chaloner, *Proposals Humbly Offered, for Passing, an Act to Prevent Clipping and Counterfeiting of Money*, pp. 4–5.

129 "opight of the Law": *Gurwoor Rodiviruv*, p. 3.

13. "HIS OLD TRICK"

130 William Paterson: Sir John Clapham, *The Bank of England*, vol. 1, p. 13, and Reginald Saw, *The Bank of England, 1694–1944*, p. 14.
under half the total needed: Reginald Saw, *The Bank of England, 1694–1944*, p. 14.

131 only forty-two members: For a good capsule account of the founding of the Bank, see Sir John Clapham, *The Bank of England*, vol. 1, pp. 13–20. Though his is something of an official history, Clapham's summary of the state of banking and credit in England immediately before the Bank's founding is also a useful overview. See also John Giuseppi's *The Bank of England*. As a long-standing Bank archivist, Giuseppi is very close to the surviving documents.

133 "marbled paper Indented": Bank of England directors' meeting minutes for August 11, 1694.
Bank of England notes: There was one earlier attempt to create bank notes, in 1661, when the Bank of Sweden issued printed notes. The Swedish experiment with paper was an immediate success, with notes trading at a premium to metal money. The Swedes—or at least their spines—had good reason to welcome the invention. Their coinage was based on cop-

per, with the result that material wealth was truly a burden. The ten-dollar coin was the heaviest piece of currency ever struck, weighing in at 19.7 kilos, or more than 43 pounds. As A. D. Mackenzie put it, "the payment of anything greater than the smallest of sums necessitated the use of heavy transport" (*The Bank of England Note*, p. 2). For all the ease that paper money brought the average Swede, though, the premium earned by paper over copper made it profitable to export Swedish coins to other markets, so that by 1664, the Bank of Sweden could not make good on its promise to exchange metal for paper on demand. With that, the notes were withdrawn, and Sweden returned to its traditional coinage as the basis for its monetary system.

August 14, 1695: Paul Hopkins and Stuart Handley, "Chaloner, William," in the *Oxford Dictionary of National Biography*.

134   brush with the turnkeys: Ibid.

14. "A Thing Impossible"

136   Neale was thoroughly overmatched: C. E. Challis, *A New History of the Royal Mint*, pp. 392–93.

137   Neale was suddenly in charge: For Neale's background, see C. E. Challis, *A New History of the Royal Mint*, pp. 392–93. For the total mint production of silver from 1660 to 1695, see Ming-hsun Li, *The Great Recoinage of 1696 to 1699*, p. 48; figures are drawn from Hopton Haynes's records. Newton's accounting for the total amount of money struck in the Great Recoinage comes from his reckoning in Mint 19.2, f. 264. The total came to £6,859,144 8s. 4d.—remarkable exactitude, but completely in keeping with Newton's passion for calculation to the limit of precision.

Official revenues disappeared: Ming-hsun Li, *The Great Recoinage of 1696 to 1699*, pp. 135–36.

138   "The people are discontented": Edmund Bohun to Hohn Cary, 31 July 1696, quoted in C. E. Challis, *A New History of the Royal Mint*, p. 387. Bohun lost his lucrative post as licensor (censor) of books when he approved the publication of a tract claiming that William and Mary held their thrones by right of their conquest of James II's forces. Fearful of even the faintest of challenges to the legitimacy of a monarchy that had in fact been seized by force of arms, Parliament ordered Bohun's arrest, questioned him at the bar of the House of Commons, burned the offending pamphlet, and stripped him of his job. There is no reason to doubt this part of his letter to Cary, but Bohun was a former supporter of James's who submitted to William's conquest a little too swiftly for both his former allies and his new masters, and it may be wise to take the extreme of his rage at the recoinage expressed elsewhere in the letter with some caution. Bohun's fate as licensor is discussed, in the larger context of the debate over the le-

gitimacy of William and Mary's claim to the monarchy, in Mark Goldie's "Edmund Bohun and *Ius Gentium* in the Revolution Debate, 1689–1693," *Historical Journal* 20, no. 3 (1977), pp. 569, 586.

the treasure of the nation was exhausted: Malcolm Gaskill, *Crime and Mentalities in Early Modern England,* p. 195.

"nobody paying or receiving": John Evelyn, diary entry for 11 June 1696, quoted in Ming-hsun Li, *The Great Recoinage of 1696 to 1699,* p. 135, and D. W. Jones, *War and Economy,* p. 137.

twenty people were arrested: Malcolm Gaskill, *Crime and Mentalities in Early Modern England,* p. 195.

"Our Coine alas it Will not Pass": The broadside most likely dates from 1697 and is quoted in full in ibid., p. 193.

139 "the new Invention of Rounding": Oath taken by Newton, 2 May 1696, *Correspondence 4,* document 548, p. 201.

No Warden had done much real work: Sir John Craig, "Isaac Newton—Crime Investigator," *Nature* 182 (1958), pp. 136, 149. For Newton's account of the organization of the Mint and the problems its antique arrangements created, see his memorandum of June 1696, *Correspondence 4,* document 552, pp. 207–9.

the post would not be too demanding: *Correspondence 4,* document 545, p. 195.

"of a prodigal temper": Mint records, cited in Richard Westfall, *Never at Rest,* p. 564.

a raise in his basic pay: Isaac Newton to the Treasury, June 1696, *Correspondence 4,* document 551, pp. 205–6.

the quality of the work of carpenters: Isaac Newton, Thomas Neale, and Thomas Hall to "the Right Honble the Lords Commissrs of his Majties Treasury," 6 May 1696, *Correspondence 4,* document 549, p. 202. (Hall was Neale's assistant.) It is reassuring to recognize that some human experiences are truly universal. Doubting one's contractor has to be one of them.

140 several needed employees: Isaac Newton and Thomas Hall to "the Right Honble the Lords Commrs. Of his Majties Treasury," 8 June 1696, *Correspondence 4,* document 550, pp. 204–5.

the grand sum of two pence: Isaac Newton to the Treasury, 1696, *Correspondence 4,* document 559, p. 218. While two pennies counted for a lot more then than now, it still amounts to more or less the cost of a large cappuccino on the streets of London today. Not much, in other words.

mastering the details of every operation: Richard Westfall describes Newton's campaign to master all the available material about the Mint in *Never at Rest,* pp. 564–66, from which this account is drawn.

"nor any other eyes than your own": Isaac Newton to All Country Mints, 16 April 1698, *Correspondence 4,* document 586, p. 271.

His holograph Mint papers: Newton's Mint papers fill Mint 19/1–5 and are held at the U.K.'s Public Record Office (Mint 19.6 is a calendar of the other five volumes). Most of his papers as Warden are in Mint 19/1, which has multiple drafts of a variety of documents associated with the recoinage, the organization of the Mint, and William Chaloner. See, for example, the nearly identical summaries of Chaloner's crimes in 19/1, f. 496, and 19/1, ff. 497–98 (the latter reproduced in *Correspondence 4*, document 581, pp. 261–62). Richard Westfall lists several other examples of Newton's need to rewrite in *Never at Rest*, p. 566, n. 47.

141 "a thing impossible": Hopton Haynes quoted in C. E. Challis, *A New History of the Royal Mint*, p. 394.

fee of twelve and three-eighths pence: Isaac Newton and Thomas Hall to the Treasury, on or after 22 February 1696/7, *Correspondence 4*, p. 236.

ten thousand pounds of refined liquid silver: C. E. Challis, *A New History of the Royal Mint*, p. 394.

142 "will hold but 700 or 650 lb wt": Isaac Newton, "Observations concerning the Mint," 1697, *Correspondence 4*, document 579, p. 256.

five hundred men and fifty horses: C. E. Challis, *A New History of the Royal Mint*, p. 394. The figure of five hundred men working at the Mint comes from Sir John Craig, *Newton at the Mint*, p. 14.

the striking chamber: Isaac Newton, "Observations concerning the Mint," 1697, *Correspondence 4*, document 579, p. 258.

"judge of the workmen's diligence": Hopton Haynes, *Brief Memoires Relating to the Silver and Gold Coins of England*, cited in Richard Westfall, *Never at Rest*, p. 561.

fifty to fifty-five times a minute: C. E. Challis, *A New History of the Royal Mint*, p. 394. He cites Haynes for Newton's calculations. The figure of fourteen men working each press comes from Isaac Newton, "Observations concerning the Mint," 1697, *Correspondence 4*, document 579, p. 258.

143 Only one man died: Richard Westfall, *Never at Rest*, p. 561.

a record output: C. E. Challis, *A New History of the Royal Mint*, p. 394.

about 2,700,000 pounds: Mint 9/60, cited in C. E. Challis, *A New History of the Royal Mint*, pp. 394, 397. See also, Ming-hsun Li, *The Great Recoinage of 1696 to 1699*, pp. 138–40. Richard Westfall dates the achievement of the 100,000-pounds-per-week output to the summer of 1696 in his *Never at Rest*, p. 561.

a peace with Louis XIV: In response to the King's complaint about a lack of cash for war and trade, in October 1696, the directors of the Bank of England were asked what could be done. Among their suggestions: "increase the species of money and expedite ye coyning thereof." See Ming-hsun Li, *The Great Recoinage of 1696 to 1699*, p. 138.

144 the enterprise would have failed: John Conduitt, notes for his biography of Newton, Keynes Ms. 130.7, 3r.

15. "THE WARDEN OF THE MINT IS A ROGUE"

147 forty pounds for each conviction: John Craig, "Isaac Newton and the Counterfeiters," *Notes and Records of the Royal Society of London* 18, no. 2 (December 1963), p. 136.

"prosecuting and swearing for money": Isaac Newton to the Treasury, July/ August 1696, *Correspondence 4*, document 553, pp. 209–10.

"Office of the Warden of his Majts. Mint": Ibid. The sentence in which Newton asked to be relieved of the duty was crossed out and rewritten in slightly different form beneath his signature: "And therefore I humbly pray that it may not be imposed upon me any longer."

149 the coining pseudonym: Paul Hopkins and Stuart Handley, "Chaloner, William," in the *Oxford Dictionary of National Biography*.

150 beyond the reach of English writs: Cooke also implicated Chaloner, without mentioning Hunter. See John Craig, "Isaac Newton—Crime Investigator," *Nature* 182, no. 4629 (July 19, 1958), p. 150.

151 Christopher Wren: As Peter Whitfield notes in his *London: A Life in Maps*, both Newgate and Bethlehem Hospital—better known as Bedlam—were rebuilt to new standards of external elegance after the Great Fire. Neither building remains; Newgate was rebuilt twice more at the same location before being demolished to provide a site for the Old Bailey law courts. Bedlam's site, in an almost too obvious irony, now boasts the Imperial War Museum.

"a kind of entrance to it": Daniel Defoe, *Moll Flanders*, p. 215.

"this abode of misery and despair": Giacomo Casanova, *The Memoirs of Jacques Casanova de Seingalt*, chapter 13, http://etext.library.adelaide.edu. au/c/casanova/c33m/chapter111.html (unabridged London edition of 1894).

152 Typhus was so widespread: The description of Newgate is drawn from Stephen Halliday's marvelous history *Newgate: London's Prototype of Hell*, pp. 30–35. I can't recommend this book too highly; it offers delightfully gruesome anecdotes in the framework of a significant case study in the evolution of prisons.

154 the Fever Islands: John Craig told the story of Newton's involvement with White and Cooke in "Isaac Newton and the Counterfeiters," *Notes and Records of the Royal Society of London* 18, no. 2 (December 1963), pp. 137–38. After his release from Newgate: *Guzman Redivivus*, p. 8.

155 Chaloner's old coining partner: Isaac Newton, "Chaloner's Case," Mint 19/1, sheet 501.

16. "BOXEFULLS OF INFORMATIONS IN HIS OWN HANDWRITING"

158 the case of the missing dies: According to John Craig's calculations, Newton appeared before the Lords Justices ten times in August and September

1696, and during that period interviewed at least six suspects, either at Newgate or at the Mint. That works out to about two days a week he devoted to the investigation, and it probably underestimates both the number of suspects questioned and the number of times each suspect was interrogated. See John Craig, "Isaac Newton and the Counterfeiters," *Notes and Records of the Royal Society of London* 18, no. 2 (December 1963), p. 137.

159 "paid Humphrey Hall to buy him a suit": Mint 19/1, sheet 467, cited in John Craig, "Isaac Newton—Crime Investigator," *Nature* 182, no. 4629 (July 19, 1958), p. 150.

160 Both men benefited: The Cambridgeshire undercover operation and George Macy's work are described in Malcolm Gaskill, *Crime and Mentalities in Early Modern England*, p. 170. The adventures of the brothers Maris and Rewse are documented in John Craig, "Isaac Newton and the Counterfeiters," *Notes and Records of the Royal Society of London* 18, no. 2 (December 1963), pp. 138–39. J. M. Beattie's *Policing and Punishment in London, 1660–1750*, pp. 228–47, details the experience of hired thief-takers in the 1690s and discusses Rewse's career in some detail.

total of £626 5s. 9d.: "An Acoumpt of monies expended by Isaac Newton of his Maj^{ts} Mint in the apprehension and prosecuting of Clippers and coyners between the third day of August 1696 and the . . . ," Mint 19/1, leaf 477.

"scandalously mercenary": Hopton Haynes, *Brief Memoires*, f. 36v, quoted in Malcolm Gaskill, *Crime and Mentalities in Early Modern England*, p. 171.

161 He used the warrant to blackmail victims: Mint 17, deposition 193, of John Holloway and Elizabeth Holloway, 14 and 17 April 1699.

hint of sexual extortion: "a certain pension," Mint 17, document 198, deposition of Mary Townsend, no date; "in company with one Mr. John Gibbons," Mint 17, document 240, deposition of Mary Townsend, 31 August 1698; "Gibbons corresponds," Mint 17, document 31, deposition of Edward Ivie, 22 August 1698. The eighty-pound price, Mint 17, document 38, deposition of Mary Hobbs, 2 July 1698; "a little adjouning room," Mint 17, document 44, deposition of Elizabeth Bond, 15 July 1698. Hints that Gibbons regularly demanded sexual payment from women—coiners in their own right and those women who came to beg or buy his intercession for relatives or lovers—come in a number of the depositions Newton took in the summer of 1698. Besides Bond's testimony about Mrs. Jackson, there are several other, similarly indirect descriptions. But explicitly, witness after witness made the point that Gibbons had his victims absolutely in his power: they could hand over whatever he sought, or they could go to Newgate and the gallows. It seems clear from reading the numerous depositions that Gibbon lusted after more than money. See also J. M. Beattie, *Policing and Punishment in England, 1660–1750*, pp. 241–42. Beattie documents the authorities' knowledge of Gibbons's untrustworthiness.

162 "sought after for coyning of Gineas and pistols": "The Examination of Elizabeth Ivie of Liccabone street in Holbourn Widdow 13 October 1698," Mint 17, document 104.

"Taverns & Prisons & other places": Isaac Newton to the Treasury, 1 October 1699, *Correspondence 4*, document 617, p. 317.

163 "boxefulls of informations" John Conduitt, *Character*, Keynes Ms. 130.7, p. 31, online at the Newton Project, http://www.newtonproject.sussex.ac.uk. England's most prolific torturing monarch: John H. Langbein, *Torture and the Law of Proof*, p. 82.

Little Ease: Ibid., p. 85. The rats did not have the desired effect on Sherwood, so his interrogation intensified with a session on the rack, ordered two weeks later.

"marching to the beat of drum": British History Online, Lambeth—Lambeth Palace, http://www.british-history.ac.uk/report.aspx?compid=45290.

164 "to cause him to be racked": David Jardine, *A reading on the use of torture in the criminal law of England before the Commonwealth*, pp. 57–58.

Archer did not speak: L. A. Parry, *The History of Torture in England*. The description of the rack comes on pages 76–77; his account of John Archer's case is on page 60. Torture continued to be used legally in Scotland long after the practice ceased in England. Most notably, in 1693 King William took advantage of the fact that he ruled two separate kingdoms to transport Henry Neville Payne from London to Edinburgh to torture him there.

evidence gained under torture: See Langbein's analysis in *Torture and the Law of Proof*, pp. 134–39.

165 "I shall have Irons put on me tomorrow": "Thomas Carter's Letter to the Warden of the Mint Sunday Afternoon," Mint 17, document 130.

"The blood of coiners and clippers": Frank Manuel, *A Portrait of Isaac Newton*, p. 244.

17. "I HAD BEEN OUT BEFORE NOW BUT FOR HIM"

168 arrested Ball and Miller: Mint 17, document 6, deposition of Mary Miller, 19 July 1698. Miller made this statement before Isaac Newton about two weeks after the events described.

Miller was a known quantity: Mint 17, document 12, deposition of Mary Miller, 5 August 1698.

169 six or seven pence per coin: Mint 17, document 27, deposition of Samuel Bond, 16 September 1698.

"imposing up on a whole Kingdom": *Guzman Redivivus*, p. 3.

171 "Debased, Diminished and Counterfeited": William Chaloner, "The Defects in the present Constitution of the Mint," p. 1.

The rot started at the top: "Report of the Committee Appointed to Inquire Into the Miscarriages of the Officers of the Mint," reprinted in

Rogers Ruding, *Annals of the Coinage of Britain and its Dependencies,* vol. 2, p. 468.

"hath got a great estate": Ibid., p. 467.

172 why not add "an Officer": William Chaloner, "The Defects in the present Constitution of the Mint," p. 1.

"Morally impossible to Counterfeit": Ibid.

173 "such bad Workmanship": Ibid.

"to perform some of his Proposals": Ibid., p. 2.

Newton's responses: Isaac Newton, untitled and undated memo, Mint 19/1, f. 496.

174 "undeniable demonstrations": "Report of the Committee Appointed to Inquire Into the Miscarriages of the Officers of the Mint," reprinted in Rogers Ruding, *Annals of the Coinage of Britain and its Dependencies,* vol. 2, p. 467.

"Mr Chaloner may make an Experiment": Mint 19/1, f. 516. Reprinted in *Correspondence 4,* pp. 231–32.

And there the matter rested: Isaac Newton, "An Answer to M^r. Chaloner's Petition" (draft), Mint 19/1, f. 499. See also John Craig, "Isaac Newton and the Counterfeiters," *Notes and Records of the Royal Society of London* 198, no. 2 (December 1963), p. 141.

Newton's testimony: "Report of the Committee Appointed to Inquire Into the Miscarriages of the Officers of the Mint," reprinted in Rogers Ruding, *Annals of the Coinage of Britain and its Dependencies,* vol. 3, pp. 533–42.

175 "libeling . . . in print": Isaac Newton, "An Answer to M^r. Chaloner's Petition" (draft), Mint 19/1, f. 499, and an untitled memo, Mint 19/1, f. 496.

18. "A New and Dangerous Way of Coining"

176 "fun[ne]d the Lords of the Treasury": "The Information of John Peers taken upon Oath ye 18th day of May 1697," Mint 17, document 86. This deposition was taken by a judge, Francis Negus, and not Newton himself. Newton heard much the same thing, possibly from Chaloner's close associate Thomas Holloway. In "An Answer to M^r. Chaloner's Petition" (draft), Mint 19/1, f. 499, Newton writes that Chaloner said he was on the verge of "funning the Parliament as he had done the King and the Bank before."

"the Character he deserv'd": *Guzman Redivivus,* p. 6.

177 "a new and dangerous way of coining": Isaac Newton, untitled memo, Mint 19/1, f. 496, and "An Answer to M^r. Chaloner's Petition" (draft), Mint 19/1, f. 499.

178 "a house in the Country": Isaac Newton, "An Answer to M^r. Chaloner's Petition" (draft), Mint 19/1, f. 499, and untitled memo, Mint 19/1, f. 496. See also "Chaloner's Case," Mint 19/1, f. 503.

179 practically indistinguishable: Isaac Newton, "Chaloner's Case," (undated,

probably late 1697), Mint 19/1, f. 503. See also parts of this account repeated in various drafts, especially in Mint 19/1, f. 496.

"might be hidden anywhere": Isaac Newton, "Chaloner's Case," (undated, probably late 1697), Mint 19/1, f. 503.

"new way quick and profitable": Ibid.

180 Chaloner make his famous boast: "The Information of John Peers taken upon Oath ye 18th day of May 1697," Mint 17, document 86.

181 He bailed his man out: He was reimbursed. See John Craig, "Isaac Newton—Crime Investigator," *Nature* 182 (1958), pp. 150–51. Newton's accounts show two sums paid out of his own funds for "frustrating the designs of Chaloner + his associates in their conspiring to set on foot a new way of coyning in summer 1697 and of apprehending some of them." The two entries total twenty-three pounds eighteen shillings, and while Newton did not itemize his expenses, the cost of funding Peers and hauling him out of jail both fit the general category here. The charges are recorded in Mint 19/1, f. 577.

Newton again arrested Holloway: Isaac Newton, "Chaloner's Case" (undated, probably late 1697), Mint 19/1, f. 503.

182 he was to arrest William Chaloner: John Craig, "Isaac Newton—Crime Investigator," *Nature* 182, no. 4629 (July 19, 1958), pp. 150–51.

Chaloner readied his counterattack: Ibid.

184 "asked him if Holloway was gone away": "The Information of Henry Saunders of Cross Lane in ye parish of St. Gyles in the Fields in the County of Middx, Tallow Chandler 25 augt. 1698," Mint 17, document 98. The details of Holloway's escape to Scotland come from this deposition and from Mint 17, documents 80, 81, and 107. Document 80 contains "The Information of rob Brown Joyner at the Turk's head near the Hermitage Bridge in Wapping 13 December 98." Documents 81 and 107 are depositions by Elizabeth Holloway.

19. "To Accuse and Vilify the Mint"

188 "such redress as shall seem best": William Chaloner, "Petition by Chaloner," late 1697, copied out by Isaac Newton, Mint 19/1, f. 497. Also *Correspondence 4,* document 580, pp. 259–60.

"raised himself by coyning": Isaac Newton, "An Answer to Mr. Chaloner's Petition," early 1698, Mint 19/1, ff. 497–98, reprinted in *Correspondence 4,* document 581, pp. 261–62.

189 "any privilege or direction was given": Ibid.

Newton was standing in the dock: John Craig, "Newton and the Counterfeiters," *Notes and Records of the Royal Society of London* 18, no. 2 (December 1963), p. 141.

gaps turned up in the plaintiff's story: *Calendar of State Papers Domestic,* 1697, p. 339.

a bald rejection: Isaac Newton, untitled and undated, Mint 19/1, f. 503.

20. "At This Rate the Nation May Be Imposed Upon"

193 "being afraid to have them": "The Examination of Thomas Carter Prisoner in Newgate 31 January 1698/9," Mint 17, deposition 118.

"they were so light": "The Deposition of John Abbot of Water Lane in Fleet street Refiner 15th day of February 1698/9," Mint 17, deposition 119.

"he would do some better": "The Examination of Thomas Carter Prisoner in Newgate 31 January 1698/9," Mint 17, deposition 118.

194 "to engage themselves in this Fund": Thomas Neale, *A Profitable Adventure to the Fortunate,* p. 2.

195 to secure the best possible return: See Anne L. Murphy's analysis in her article "Lotteries in the 1690s: Investment or Gamble?," *Financial History Review* 12, no. 2 (2006), pp. 231–32.

a quarter of a million pounds behind: Ibid.

"Credit and Honour of the Nation": Anonymous petitioner quoted in ibid., p. 231.

196 actual bills of exchange: Thomas Neale, "Fourteen Hundred Thousand Pound, made into One Hundred and Forty Thousand Bills of Ten Pounds apiece, to be given out for so much as Occasion requires, and to be paid as Chance shall Determine in course, out of 1515000 l. being left to be only made use of to pay Interest, Premium and Charge," 1697.

"nobody does or will understand": Anne L. Murphy, "Lotteries in the 1690s: Investment or Gamble?," *Financial History Review* 12, no. 2 (2006), p. 233.

forty-five thousand pounds' worth: "The fourth parliament of King William: First Session," *History and Proceedings of the House of Commons,* vol. 3, pp. 91–106, http://www.british-history.ac.uk/report.aspx?compid=37657.

202 the smaller and less valuable fish: All the quotes and the narrative of Davis's pursuit of Chaloner come from the deposition in which Davis described the information he provided Secretary Vernon, Mint 17, document 85.

Robert Morris: John Craig, "Newton and the Counterfeiters," *Notes and Records of the Royal Society of London* 18, no. 2 (December 1963), p. 142.

21. "He Had Got His Business Done"

204 "why am I so strictly confined": William Chaloner, "Letter to the Warden of the Mint," undated, Mint 17, document 133.

"he had a Trick left yet": *Guzman Redivivus,* p. 7.

205 Brady had received some of his supply: "The Examination of Edward Ivy

als Jonas late of [*sic*] in London Gentl[eman] taken before me the 17th day of May 1698," Mint 17, document 31.

Newgate or some other jail: See, for example, Mint 17, document 99, "The Information of Mary Townsend of the p[ar]ish of St. Andrews Holborn in the County of Midd[lese]x Widdow taken this 31 day of Augt. 1698," a deposition in which Townsend implicates John Gibbons and Edward Ivy. Two months later, Elizabeth Ivy complains of Gibbons and John Jennings in document 104, along with Brady—all of whom were named later by her husband—and someone Townsend omitted, a Valentine Cogswell. Newton already knew about Cogswell, though, because in May, an informant whose name is missing from the surviving portion of his or her deposition mentioned "Cogswell a Gent[leman]," whom Brady initiated into the coiners' society. That witness was clearly well placed, listing eighteen men and women involved in clipping, coining, and other crimes in document 91—a tally that included the horrifying charge "that one Capt. Tuthill" kept "a Rape Mill." This informant *did* mention Chaloner under his pseudonym Chandler, reporting that his coining operation ran under the protection of John Gibbons—but here again, his was just one name in a list. The point of the testimony as recorded by Newton's clerk was to paint a synoptic picture of a capital overrun with counterfeiters from every class.

206 more than 140: The Mint 17 file contains obvious omissions—at one point there are some one hundred numbers missing in the document sequence. I share the conclusion of John Craig ("Isaac Newton—Crime Investigator," p. 151), that up to half of the relevant documents are missing.

Newton and his witness signed the document: This procedure is pretty clear from the depositions themselves, but I was pointed to this view of Newton's handling of the witnesses by John Craig in "Isaac Newton—Crime Investigator," p. 151.

"abot. 7 or 8 years ago she hath seen": "The Information of Katherine Coffee wife of Patrick Coffee Goldsmith late of Aldermanbury by Woodstr[eet] 18 day of February 1698/9," Mint 17, document 124.

207 "abot 10 of thes counterfeit Gineas": "The Information of Katherine Matthews [aka Carter] of Earles Cour in Cranborn ally in the p[ar]ish of St. Anns Westm[inste]r," Mint 17, document 116.

"either Ginea Dyes or half Crown Dyes": "The Depostion of Humphrey Hanwell of Lambeth p[a]rish in Southwark 22d Feb 1698/9," Mint 17, document 123.

a document he titled "Chaloner Case": "Chaloner's Case," Mint 19/1, ff. 501–4.

208 six hundred pounds of false half-crowns: "The Deposition of John Abbot of Water Lane in Fleet street Refiner 15th. day of February 1698/9," Mint 17, document 119.

"a Coyning Press at Chiswick": "The Information of Cecilia Labree 6ᵗʰ Feby 1698/9," Mint 17, document 143.

22. "If Sᴿ Be Pleased . . ."

209 "joined they should save themselves": "The Examination of Thomas Carter Prisoner in Newgate 31 January 1698/9," Mint 17, document 118.
seven times in the pillory: "Chaloner's letter to Mr. Secretary Vernon," Mint 17, document 126.

210 "an evidence agt. him": "Carter's Letter to Mr Secretary Vernon," undated, Mint 17, document 130. Carter's visits to the pillory included one for forging a Bank of England note in 1696, lending a powerful tinge of authority to Chaloner's hope of discrediting Carter before a jury. See *The Proceedings of the Old Bailey* for trials completed on 9 December 1696, http://www .hrionline.ac.uk/oldbailey/html_units/1690s/t16961209-59.html.
Newton's good graces: Thomas Carter in Mint 17, documents 83, 84, 118, 123, and 130.
"if ever King James came again": "The Deposition of Samuel Bond of Ashbourn in the parke in the Count of Derby Chyrugeon 16 October, 1698," Mint 17, document 27s.
"to come to the Dogg": "John Whitfield's Lettr to the Isaac Newton Esqʳ Warden of His Majᵗʸˢ Mint Febry 9ᵗʰ 98/9," Mint 17, document 134.

211 "such vacan[t] places": Ibid.
"Chaloner was a little suspicious": "Thomas Carter's Letter to the Warden of the Mint Sunday afternoon," undated, Mint 17, document 130.
"All his discourse to him": Ibid.
"I will get out of him": Ibid.

212 "20 such as the Warden": "The Information of William Johnson Farrier at the Barbers Pole near the Watchhouse in Radcliff highway taken this 8ᵗʰ day of February 1698/9," Mint 17, document 145. See also document 146, "The Information of Ann Duncomb of Black cheek ally in East Smithfield Spinst[er] 8 Feb: 1698/9," and document 148, "The Examination of the Josiah Cook of Eagle Str[eet] of St Gyles in the fields in the County of Midd[lese]x Chyrurgeon 14 February 1698/9," for more of what Newton could hold over Lawson's head.
"I hope Charity will moe": "Letter sent to Is[aac] Newton Esqr. From John Ignatius Lawson Sunday night and Munday morning," undated, Mint 17, document 132, and "The Information of John Ignatius Lawson Vizt," 3 April 1699, Mint 17, document 199.
Lawson kept such tales coming: Each of the separate incidents in this paragraph are contained in "The Information of John Ignatius Lawson Vizt," 3 April 1699, Mint 17, document 199. (This document number contains a series of reports Lawson sent to Newton between January and April 1699.)

213 "the case of a small Bible": Ibid.

so he slipped through: "The Information of Jno Ignatius Lawson now Prisoner in Newgate 3ᵈ. Aprill 1699," Mint 17, document 165.

214 "before she confess any thing": "John Ignatius Lawson's Letter to Is: Newton Esqr," undated, Mint 17, document 131.

Elizabeth Holloway: "Letter sent to Is[aac] Newton Esqr. From John Ignatius Lawson Sunday night and Munday morning," undated, Mint 17, document 132.

"hath seen him coyn": Ibid.

"no man in England could grave": "The Information of Jno Ignatius Lawson now Prisoner in Newgate 3ᵈ. Aprill 1699," Mint 17, document 165. The date is almost certainly a reference to the day when Newton collated the information Lawson provided him (see the following note), as this testimony appears in sequence with depositions taken in February, and the information it contains implies a date before the March 1 start of the Old Bailey sessions for which Newton was preparing his case.

215 "Tin plated over with Silver": John Ignatius Lawson, untitled report, 25 February 1698, countersigned by Newton on 3 April 1699, Mint 17, document 184.

deceit that had enraged Elizabeth Holloway: John Ignatius Lawson, untitled and undated report, Mint 17, document 192.

"6 of one Jury and 8 of another": "Letter sent to Is[aac] Newton Esqr. From John Ignatius Lawson Sunday night and Munday morning," undated, Mint 17, document 132.

216 "the best accᵗ I cann remember": "William Chaloner Letter to the Warden of the Mint," Mint 17, document 133.

23. "IF I DIE I AM MURTHERED"

217 "Some p[e]rsons agᵗ my desire": "William Chaloner Letter to the Warden of the Mint," Mint 17, document 133, first letter.

219 "I am murthered": "William Chaloner Letter to the Warden of the Mint," Mint 17, document 133, second letter.

"suggestions of such evill persons": "A Copy of a Letter directed from Will Chaloner to Justice Railton," Mint 17, document 133, fourth letter.

220 "in flatt stitch": "William Chaloner Letter to the Warden of the Mint," Mint 17, document 133, third letter.

"Coyn or paper": John Ignatius Lawson, untitled and undated report, Mint 17, document 192.

221 "wasted and spoiled": "William Chaloner's Letter to Isaac Newton Esq," Mint 17, document 174.

222 "5 times before": "John Ignatius Lawson's Letter to Is: Newton Esqr," undated, Mint 17, document 131.

"such frightful Whimseys": *Guzman Redivivus*, p. 7.

"pretend himself sick": John Ignatius Lawson, untitled report, 25 February 1698, countersigned by Newton on 3 April 1699, Mint 17, document 184.

"counterfeiting the Madman": *Guzman Redivivus*, p. 7.

24. "A Plain and Honest Defence"

224 indiscriminately silver and gold: I take this point from, and first found the details of the indictment in, John Craig, "Isaac Newton—Crime Investigator," *Nature* 182 (1958), p. 151. Craig led me to the Middlesex Sessions Roll for 1699, now in the London Metropolitan Archive, from which he drew the account I repeat here.

225 an admission of guilt: J. M. Beattie, *Crime and the Courts in England, 1660–1800*, p. 337.

"a plain and honest Defence": William Hawkins, *A Treatise of the Pleas of the Crown*, vol. 2, quoted in J. M. Beattie, *Policing and Punishment in London, 1660–1750*, p. 264.

226 criminals yet uncaught: See "History of the Old Bailey Courthouse," *The Proceedings of the Old Bailey*, http://www.oldbaileyonline.org/history/the-old-bailey/.

judges "should be the advocat": newspaper commentary from 1783, quoted in J. M. Beattie, *Crime and the Courts in England, 1660–1800*, p. 345.

228 worse judge: This account of Lovell's career and character has been drawn from Tim Wales's entry "Lovell, Sir Salathiel" in the *Oxford Dictionary of National Biography*. The history of Lovell's role in recommending pardons is slightly more complicated than the gloss above. As Wales writes, Lovell's arbitrary (and under suspicion of corruption) use of power to tip the balance toward or away from clemency embroiled him in disputes with London's aldermen at least twice, and he did not maintain a completely free hand. In 1699, however, he was between battles and retained significant power in this area.

calling the defendant notorious: See John Craig, "Isaac Newton and the Counterfeiters," *Notes and Records of the Royal Society of London* 18 (1963), p. 142. Unfortunately, no more vivid account of Lovell's invective exists; the surviving documents detail only the indictments, the witness list, and some of the evidence presented at trial. The records for the March 1699 quarter sessions of the witnesses' statements and other direct records of what was said at trials have been lost.

229 "Gineas which were reputed Chaloners": "The Information of Katherine Coffee wife of Patrick Coffee Goldsmith late of Aldermanbury by Woodstr[eet] 18 day of February 1698/9," Mint 17, document 124.

the prisoner's hammer: Taylor's role in producing both pistole and guinea dies survives in several hearsay depositions; see, for example, Katherine

Carter's testimony before Isaac Newton on 21 February 1698/9, Mint 17, document 122. Newton amassed enough testimony to state with apparent certainty in his summary document "Chaloner's Case" that Taylor had produced two sets of French dies in 1690 and 1691, and one for guineas in 1692. See Mint 19/1.

Both were almost certainly lying: John Craig, "Isaac Newton and the Counterfeiters," p. 143.

230 "a Treat at the 3 Tuns": "The Deposition of John Abbot of Water Lane in Fleet street Refiner 15th. day of February 1698/9," Mint 17, document 119.

231 "affronting Mr. Recorder": *Guzman Redivivus*, p. 10.
by the midday meal break: This data comes from the December 1678 session of the Old Bailey, reported in J. M. Beattie, *Policing and Punishment in London, 1660–1750*, p. 262.

232 Lawson walked out of court: *Proceedings of the Old Bailey*, 11 October 1699, p. 3. Online at http://www.oldbaileyonline.org/images.jsp?doc16991011003. "The Evidence was plain": Ibid.
"Guilty of High-Treason": *Guzman Redivivus*, p. 11.

25. "O I Hope God Will Move Yoʀ Heart"

233 "a Harping Ruin": *Guzman Redivivus*, p. 12.
"in[no]sent he was": "Carter's Letter to Is: Newton Esq," Mint 17, document 130.

234 "Yo[u]r near murdered humble Servt": "William Chaloner's Letter to Isaac Newton Esq.," Mint 17, document 205.

235 a list of complaints: *Guzman Redivivus*, p. 12.
the condemned men's pew: David Kerr Cameron, *London's Pleasures*, p. 144.
"Charity and Forgiveness": *Guzman Redivivus*, p. 12.

236 "hanged by the neck": See V.A.C. Gatrell, *The Hanging Tree*, pp. 315–16.

237 "a rotten Member": *Guzman Redivivus*, p. 13.

Epilogue

238 under "this gentleman's care": Hopton Haynes, *Brief Memoirs*, quoted in G. Findlay Shirras and J. H. Craig, "Sir Isaac Newton and the Currency," *Economic Journal* 55, no. 218/219, p. 229.

239 £1,650 a year: Richard Westfall, *Never at Rest*, pp. 606–7. Westfall notes that early in Newton's mastership, the resumption of war with France bit into the supply of gold and silver to the Mint, and hence Newton's profit on coining operations. It was a feast-or-famine job, but Newton was able to hold on to it long enough to grow legitimately wealthy and to live precisely as he pleased.
"To explain all nature": Cambridge Add. Ms. 3790.3, f. 479, quoted in Robert Westfall, *Never at Rest*, p. 643.

240 "Nature is very constant": *Opticks,* from query 31, added in the Latin
edition and republished in every subsequent version. Quoted in Robert
Westfall, *Never at Rest,* p. 644.

"Is not infinite Space": From the Latin *Optice* and the second edition
Opticks, query 28, with edits to the original translation from the Latin by
Robert Westfall, *Never at Rest,* p. 647.

241 true nature of Christ's body: For a discussion of Newton's speculation about
the body of Christ, see Betty Jo Teeter Dobbs, *The Janus Faces of Genius,*
pp. 214–15. The question of angelic prospects after the Apocalypse is discussed on p. 32 of "Newtonian Angels," a draft chapter of an upcoming
book by Simon Schaffer. Newton worked on prophetic chronology for the
better part of two decades, coming up with several different possible dates
for the second coming. See Robert Westfall, *Never at Rest,* pp. 815–17.

"strong meats for men": The phrase occurs in Newton's draft of a massive
project on the history of the church from its origins in the early centuries of the Christian era, cited by Simon Schaffer on p. 33 of "Newtonian
Angels."

most of the new silver: G. Findlay Shirras and J. H. Craig, "Sir Isaac Newton and the Currency," *Economic Journal* 55, no. 218/219, p. 229.

That imbalance sucked silver: G. Findlay Shirras and J. H. Craig, "Sir Isaac
Newton and the Currency," *Economic Journal* 55, no. 218/219, pp. 228–36.

242 even his old currency ally: For Newton's argument and Lowndes's reply,
see ibid., pp. 230–31.

243 some historians have credited: See, for example, Fernand Braudel, *The
Wheels of Commerce,* pp. 525–28.

244 He bought more: The sequence of Newton's involvement in the South Sea
Company is detailed in Richard Westfall, *Never at Rest,* pp. 861–62.

245 the value of his estate: Ibid., pp. 862, 870.

"that he could not calculate": Reported in Joseph Spence, *Anecdotes, Observations, and Characters, of Books and Men,* p. 368, cited in Robert Westfall,
Never at Rest, p. 862. The classic account of the South Sea Bubble remains Charles Mackay's *Extraordinary Popular Delusions and the Madness
of Crowds,* pp. 55–104. Harvard Business School houses a major collection
of material on the bubble, and its website offers a good brief introduction
to that remarkable year: http://www.library.hbs.edu/hc/ssb/index.html.

he lived moderately: Robert Westfall, *Never at Rest,* p. 850.

"He generally made a present": William Stukeley, *Memoirs of Sir Isaac
Newton's Life,* pp. 68–69, cited in Richard Westfall, *Never at Rest,* p. 856.
The original recollection comes in a letter Stukeley wrote to Richard
Mead shortly after Newton's death, on 15 July 1727, in Keynes Ms. 136,
part 3, p. 11, and online at the Newton Project, http://www.newtonproject
.sussex.ac.uk/texts/viewtext.php?id=THEM00158&mode=normalized.

active interest in the Royal Society: See Richard Wcstfall, *Never at Rest,* pp. 863–64.

246 "a little way of in the country": The recollection of James Stirling, cited in ibid., p. 866.

a visitor came to call: See Zachary Pearce's recollection of his visit with Newton in ibid., p. 869.

The illness persisted: For a fuller account of Newton's last days, see ibid., pp. 863–70.

"only like a boy playing": Reported in Joseph Spence, *Anecdotes, Observations, and Characters, of Books and Men,* p. 54. It has been repeated throughout the Newton literature and is, *inter alia,* reprinted in *Never at Rest,* p. 863.

247 "If you doubt there was such a man": The seventeenth- and eighteenth-century view of Newton as an angelic figure, an intermediary between God and the created universe, is beautifully analyzed in Simon Schaffer's draft chapter "Newtonian Angels"—in its entirety for the larger question, pp. 8–9 for the Fatio-Conduitt communication.

Bibliography

NEWTON'S CORRESPONDENCE AND MANUSCRIPTS

Almost all of the known letters to and from Isaac Newton have been collected in *The Correspondence of Isaac Newton*. Seven volumes, edited by H. W. Turnbull, J. F. Scott, A. R. Hall, and Laura Tilling. Cambridge: Cambridge University Press for the Royal Society, 1959–1977. (A few letters still turn up, including, recently, a handful between Newton and Fatio exchanged long after the break in their relationship in 1693.)

Newton's manuscripts are widely scattered. For this book the most important archive is held at the National Archives, in Kew, England. Newton's holograph Mint records are in six folios, Mint 19/1–6. The depositions taken in his presence are collected in Mint 17.

Other important locations for Newton documents include Kings College, Cambridge; the Cambridge University Library; the Jewish National and University Library in Jerusalem (home of many of Newton's theological papers); the Bodleian Library and the Burndy Library collection, now housed at the Huntington Library in Pasadena. I tracked these collections bibliographically and online, and consulted the documents housed in them as needed for this book through two main off-site routes.

The first was through the work of the Newton Project, which can be found at http://www.newtonproject.sussex.ac.uk/prism.php?id=1. The project has published a wide-ranging selection of Newton's original writings, adding translations as necessary. Of special value to me were the transcriptions of all of Newton's surviving early notebooks. The collection also offers a very valuable set of accounts of Newton by contemporary or near-contemporary observers. Several of Newton's reports as Master of the Mint of significance to the conversion from a silver to a gold standard are available online at http://www.pierre-marteau.com/editions/1701-25-mint-reports.html#masters.

Last, the Harvard University Library holds a copy of the hard-to-find Chadwyck Healy microfilm edition of *Sir Isaac Newton, 1642–1727: Manuscripts and Papers*, edited with a finding aid written by Peter Jones: *Sir Isaac Newton: A Catalogue of Manuscripts and Papers*. The edition contains photographs of the bulk of Newton's manuscript output from 1660 on. It is not quite complete, but it is the nearest thing to a comprehensive collection in existence. It is not what you would call easy to use, as the quality of the photographs varies enormously, but I found it an invaluable resource.

Newton's Published Books

The Principia: Mathematical Principles of Natural Philosophy. Translated by
I. Bernard Cohen and Anne Whitman, assisted by Julia Budenz.
Berkeley: University of California Press, 1999.
Opticks. New York: Dover, 1952 (preface by I. B. Cohen, c. 1979).

The Cohen and Whitman edition of the *Principia* remains the definitive choice for readers of English for three reasons. The translation itself is admirably clear and transparent to Newton's argument; the design of the edition does everything it can to make this dense material as easy to follow as possible; and above all, Cohen's guide to Newton's text, a book-length work in its own right, is invaluable. Other editions have come out since, but accept no substitutes. My copy of *Opticks* is the one with Albert Einstein's charming, brief tribute to Newton.

Biographies

Brewster, Sir David. *The Life of Sir Isaac Newton,* revised and edited by
W. T. Lynn. London: Gall & Inglis, 1875.
Craig, Sir John. *Newton at the Mint.* Cambridge. Cambridge University
Press, 1946.
Fara, Patricia. *Newton: The Making of a Genius.* New York: Columbia University Press, 2002.
Gleick, James. *Isaac Newton.* New York: Pantheon Books, 2003.
Hall, A. Rupert. *Isaac Newton: Adventurer in Thought.* Oxford: Blackwell
Publishers, 1992.
Manuel, Frank. *A Portrait of Isaac Newton.* Washington, D.C.: New Republic Books, 1979.
Westfall, Richard S. *Never at Rest.* Cambridge: Cambridge University
Press, 1980.
White, Michael. *Isaac Newton: The Last Sorcerer.* New York: Basic Books,
1999.

At different times in the course of this project I consulted a wide range of biographies. Like most writers on Newton since 1980, I am most deeply indebted to Richard Westfall's scholarly, accessible, and comprehensive biography. Westfall is one of the giants without whom this book could not have been written. The best short introduction to Newton's life and work is James Gleick's brief life. It is beautifully written and provides a very clear account of what it was that Newton did that makes him still so important; Gleick also manages to convey the context of Newton's life and times in an extremely concise package. Frank Manuel's *Portrait* was the book that got me started on this project; in it he quotes Chaloner's last letter to Newton, and when I first read it, almost twenty years ago, it left a question—What on earth was Newton doing in contact with a condemned coiner?—that this book attempts to answer. Craig's *Newton at the Mint* is the only book-length study of that period of Newton's life; it touches on Newton's tenure as Warden only briefly, but still, it's all there was. Fara's and Hall's works are aimed more at a professional audience than the lay public; both are full of valuable insights. Brewster's massive account is as much a historical document—an illustration of Victorian priorities—as it is a currently useful account of Newton. I don't always agree with Michael White's emphases, but it was the first popular Newton biography that I'm aware of to focus on what has been of scholarly interest for some time—the connection between the long-ignored history of Newton's alchemical work and his more "respectable" interests in what we now call science.

OTHER SOURCES

Abrahamson, Daniel M. *Building the Bank of England.* New Haven: Yale University Press, 2005.

Anderson, Michael, ed. *British Population History.* Cambridge: Cambridge University Press, 1996.

Anonymous. *Guzman Redivivus: A Short View of the Life of Will. Chaloner.* London: printed for J. Hayns, 1699.

Appleby, Joyce Oldham. "Locke, Liberalism and the Natural Law of Money." *Past and Present,* no. 71 (May 1976), pp. 43–69.

Axtell, James I. "Locke's Review of the *Principia.*" *Notes and Records of the Royal Society of London* 20, no. 2 (1965), pp. 152–61.

Barter, Sarah. "The Mint." In John Charlton, ed., *The Tower of London.* London: HMSO, 1978.

Beattie, J. M. *Crime and the Courts in England, 1660–1800.* Oxford: Clarendon Press, 1986.

———. *Policing and Punishment in London, 1660–1750.* Oxford: Oxford University Press, 2001.

———. "The Cabinet and the Management of Death at Tyburn after the Revolution of 1688–1689." In Lois G. Schwoerer, ed., *The Revolution of 1688–1689.* Cambridge: Cambridge University Press, 1992.

Braudel, Fernand. *Civilization and Capitalism.* Volume 2: *The Wheels of Commerce.* New York: Harper and Row, 1982.

Bricker, Phillip, and R.I.G. Hughes. *Philosophical Perspectives on Newtonian Science.* Cambridge, Mass.: MIT Press, 1990.

Brown, E. H. Phelps, and Sheila V. Hopkins. "Seven Centuries of the Prices of Consumables Compared with Builders' Wage-Rates." *Economica* 23, no. 92, new series (November 1956), pp. 296–314.

Byrne, Richard. *The London Dungeon Book of Crime and Punishment.* London: Little, Brown, 1993.

Cameron, David Kerr. *London's Pleasures.* Stroud, Gloucestershire: Sutton Publishing, 2001.

Challis, C. E., ed. *A New History of the Royal Mint.* Cambridge: Cambridge University Press, 1992.

Chaloner, William. *Proposals Humbly offered, for Passing an Act to prevent Clipping and counterfeiting of Mony.* London, 1694.

———. "The Defects in the present Constitution of the Mint." London, 1697; British Library ascription, 1693.

Chandrasekhar, S. *Newton's* Principia *for the Common Reader.* Oxford: Clarendon Press, 1995.

Charlton, John, ed. *The Tower of London: Its Buildings and Institutions.* London: HMSO, 1978.

Childs, John. *The Nine Years' War and the British Army, 1688–1697.* Manchester: Manchester University Press, 1991.

Clapham, Sir John. *The Bank of England.* Two volumes. Cambridge: Cambridge University Press, 1970.

Clark, William, Jan Golinski, and Simon Schaffer, eds. *The Sciences in Enlightened Europe.* Chicago: University of Chicago Press, 1999.

Clarke, Desmond. *Descartes: A Biography.* Cambridge: Cambridge University Press, 2006.

Claydon, Tony. *William III.* London: Longman, 2002.

Cohen, I. Bernard, and George E. Smit, eds. *The Cambridge Companion to Newton.* Cambridge: Cambridge University Press, 2002.

Cohen, I. Bernard, and Richard S. Westfall, eds. *Newton.* New York: W. W. Norton, 1995.

Coleman, David, and John Salt. *The British Population.* Oxford: Oxford University Press, 1992.

Cook, Alan. *Edmond Halley.* Oxford: Clarendon Press, 1998.

Cragg, Gerald R. *Reason and Authority in the Eighteenth Century.* Cambridge: Cambridge University Press, 1964.

Craig, Sir John. "Isaac Newton and the Counterfeiters." *Notes and Records of the Royal Society of London* 18 (1963), pp. 136–45.

———. "Isaac Newton—Crime Investigator." *Nature* 182 (1958), pp. 149–52.

Cranston, Maurice. *John Locke: A Biography*. Oxford: Oxford University Press, 1985.

DeBeer, E. S., ed. *The Correspondence of John Locke*. Volume 4. Oxford: Oxford University Press, 1979.

Defoe, Daniel. *A Journal of the Plague Year*. Oxford: Oxford University Press, 1990.

———. *Moll Flanders*. New York: W. W. Norton, 2004.

Dickson, P.G.M. *The Financial Revolution in England*. Aldershot, Hampshire: Gregg Revivals, 1993.

Dobbs, Betty Jo Teeter. *Alchemical Death and Resurrection*. Washington, D.C.: Smithsonian Institution Libraries, 1990.

———. *The Janus Faces of Genius: The Role of Alchemy in Newton's Thought*. Cambridge: Cambridge University Press, 1991.

———, and Margaret C. Jacob. *Newton and the Culture of Newtonianism*. Atlantic Highlands, N.J.: Humanities Press, 1995.

Earle, Peter. *A City Full of People: Men and Women of London, 1650–1750*. London: Methuen, 1994.

Fauvel, John, et al., eds. *Let Newton Be!* Oxford: Oxford University Press, 1988.

Fay, C. R., "Newton and the Gold Standard." *Cambridge Historical Journal* 5, no. 1 (1935), pp. 109–17.

Feingold, Mordechai. *The Newtonian Moment*. New York: New York Public Library/Oxford University Press, 2004.

Force, James E., and Sarah Hutton, eds. *Newton and Newtonianism*. Dordrecht, Netherlands: Kluwer Academic Publishers, 2004.

Gaskill, Malcolm. *Crime and Mentalities in Early Modern England*. Cambridge: Cambridge University Press, 2000.

Gatrell, V.A.C. *The Hanging Tree*. Oxford: Oxford University Press, 1994.

Gerhold, Dorian. "The Growth of the London Carrying Trade, 1681–1838." *Economic History Review* 41, no. 3, 2nd series (1988), pp. 392–410.

Giuseppi, John. *The Bank of England*. London: Evan Brothers, 1966.

Goldie, Mark. "Edmund Bohun and *Ius Gentium* in the Revolution Debate, 1689–1693." *Historical Journal* 20, no. 3 (1977), pp. 569–86.

Golinski, Jan. *British Weather and the Climate of Enlightenment*. Chicago: University of Chicago Press, 2007.

Goodstein, David L., and Judith R. Goodstein. *Feynman's Lost Lecture: The Motion of Planets Around the Sun*. New York: W. W. Norton, 2000.

Green, Nick, ed. *The Bloody Register*. Volume 1. London: Routledge/Thoemmes Press, 1999.

Guerlac, Henry. *Newton on the Continent*. Ithaca, N.Y.: Cornell University Press, 1981.

Hall, A. Rupert. *Isaac Newton: Eighteenth-Century Perspectives*. Oxford: Oxford University Press, 1999.

———. *Newton: His Friends and His Foes*. Aldershot, Hampshire: Variorum, 1993.

Halliday, Stephen. *Newgate: London's Prototype of Hell*. Stroud, Gloucestershire: Sutton Publishing, 2006.

Harding, Christopher, et al. *Imprisonment in England and Wales*. London: Croom Helm, 1985.

Harman, P. M., and Alan E. Shapiro, eds. *The Investigation of Difficult Things*. Cambridge: Cambridge University Press, 1992.

Harrison, John. *The Library of Isaac Newton*. Cambridge: Cambridge University Press. 1978.

Hayward, Arthur L., ed. *Lives of the Most Remarkable Criminals*. New York: Dodd, Mead, 1927.

Heal, Felicity, and Clive Holmes. *The Gentry in England and Wales, 1500–1700*. London: Macmillan, 1994.

Herivel, John. *The Background to Newton's Principia*. Oxford: Clarendon Press, 1965.

Herrup, Cynthia B. *The Common Peace*. Cambridge: Cambridge University Press, 1987.

Heyd, Michael. *"Be Sober and Reasonable": The Critique of Enthusiasm in the Seventeenth and Eighteenth Centuries*. Leyden, Netherlands: Brill, 1995.

Hopkins, Paul, and Stuart Handley. "Chaloner, William." Entry in the *Oxford Dictionary of National Biography*, edited by H.C.G. Matthew and Brian Harrison. Oxford: Oxford University Press, 2004.

Horsefield, J. Keith. "Inflation and Deflation in 1694–1696." *Economica* 23, no. 91, new series (August 1956), pp. 229–43.

Horton, S. Dana. *The Silver Pound*. London: Macmillan, 1887.

Houghton, Thomas. *Proposals for a Fund of A Hundred and Fifty Thousand Pounds per Annum*. London, 1694.

Hunter, Michael, ed. *Robert Boyle Reconsidered*. Cambridge: Cambridge University Press, 1994.

Hurl-Eamon, Jennine. "The Westminster Impostors: Impersonating Law Enforcement in Early-Eighteenth-Century London." *Eighteenth Century Studies* 38, no. 3 (2006), pp. 461–83.

Inwood, Stephen. *A History of London*. London: Macmillan, 1998.

Jardine, David. *A reading on the use of torture in the criminal law of England before the Commonwealth*. London: Baldwin and Craddock, 1837.

Jones, D. W. *War and Economy in the Age of William III and Marlborough*. Oxford: Basil Blackwell, 1988.

Jones, J. R. *The Revolution of 1688 in England.* London: Weidenfeld and Nicolson, 1972.

Jonson, Ben. *The Alchemist.* http://www.gutenberg.org, 10th edition, May 2003.

King-Hele, D. G., and A. R. Hall, eds. *Newton's* Principia *and Its Legacy.* London: The Royal Society, 1988.

Koyré, Alexandre. *Newtonian Studies.* Cambridge, Mass.: Harvard University Press, 1965.

Lander, J. "Burial Seasonality and Causes of Death in London, 1670–1819." *Population Studies* 42, no. 1 (March 1988), pp. 59–83.

———. "Mortality and Metropolis: The Case of London, 1675–1825." *Population Studies* 41, no. 1 (March 1987), pp. 59–76.

Langbein, John H. *Torture and the Law of Proof.* Chicago: University of Chicago Press, 1977.

Laundau, Norma, ed. *Law, Crime and English Society, 1660–1830.* Cambridge: Cambridge University Press, 2002.

Li, Ming-hsun. *The Great Recoinage of 1696 to 1699.* London: Weidenfeld and Nicolson, 1963.

Library of the House of Commons. "Inflation: The Value of the Pound, 1750–2002." Research Paper 02/82, 11 November 2003.

Linebaugh, Peter. *The London Hanged: Crime and Civil Society in the Eighteenth Century.* London: Verso, 2003.

Liss, David. *A Conspiracy of Paper.* New York: Random House, 2000.

Locke, John. *The Works of John Locke.* Ten volumes. Darmstadt, Germany: Scientia Verlag, 1963.

Lodge, Sir Richard. *The Political History of England.* London: Longmans Green, 1923.

Luttrell, Narcissus. *A Brief History of State Affairs.* Six volumes. Oxford: Oxford University Press, 1857.

Macaulay, Lord. *The History of England.* Six volumes. Edited by Charles Harding Firth. New York: AMS Press, 1968.

Macfarlane, Alan. *Justice and the Mare's Ale.* Cambridge: Cambridge University Press, 1981.

Mackay, Charles. *Extraordinary Popular Delusions and the Madness of Crowds.* Petersfield, Hampshire: Harriman House, 2003.

Mackenzie, A. D. *The Bank of England Note.* Cambridge: Cambridge University Press, 1953.

Mayhew, Nicholas. *Sterling: The History of a Currency.* New York: John Wiley, 1999.

McGuire, J. E., and P. M. Rattansi. "Newton and the Pipes of Pan." *Notes and Records of the Royal Society of London* 21 (1966), pp. 108–48.

McGuire, J. E., and Martin Tamny. *Certain Philosophical Questions: Newton's Trinity Notebook.* Cambridge: Cambridge University Press, 1983.

McKay, Derek, and H. M. Scott. *The Rise of the Great Powers, 1648–1714.* London: Longman Group, 1983.

McKie, D., and G. R. de Beer. "Newton's Apple." *Notes and Records of the Royal Society of London* 9, no. 1 (October 1951), pp. 46–54, and no. 2 (May 1952), pp. 333–35.

McLynn, Frank. *Crime and Punishment in Eighteenth-Century England.* London: Routledge, 1989.

McMullan, John L. *The Canting Crew: London's Criminal Underworld, 1550–1700.* New Brunswick, N.J.: Rutgers University Press, 1984.

More, Louis Trenchard. *The Life and Works of the Honorable Robert Boyle.* Oxford: Oxford University Press, 1944.

Murphy, Anne L. "Lotteries in the 1690s: Investment or Gamble?" *Financial History Review* 12, no. 2 (2006), pp. 227–46.

Murphy, Theresa. *The Old Bailey.* Edinburgh: Mainstream Publishing, 1999.

Neale, Thomas. *A Profitable Adventure to the Fortunate.* London: F. Collins in the Old-Bailey, 1694.

———. "Fourteen Hundred Thousand Pound, made into One Hundred and Forty Thousand Bills of Ten Pounds apiece, to be given out for so much as Occasion requires, and to be paid as Chance shall Determine in course, out of 1515000 l. being left to be only made use of to pay Interest, Premium and Charge." 1697. Location: British Library.

Nelson, Sarah Jones. Unpublished essay on Newton, Locke, and the axiomatic foundations of natural rights. Private communication, 2004.

Newman, William R. *Atoms and Alchemy.* Chicago: University of Chicago Press, 2006.

———, and Lawrence Principe. *Alchemy Tried in the Fire.* Chicago: University of Chicago Press, 2002.

Newton, Humphrey. Keynes Ms. 135 (two letters from Humphrey Newton to John Conduitt, 17 January 1727/8 and 14 February 1727/8). Cambridge: King's College, Cambridge University.

Palter, Robert, ed. *The Annus Mirabilis of Sir Isaac Newton, 1666–1696.* Cambridge, Mass.: MIT Press, 1970.

Parkhurst, Tho[mas]. *The True Way of Taxing shewing What is the Legal Rack-Rent for Taxing first of Laymen, secondly of Churchmens Real Estates Equally.* London, 1693.

Parry, L. A. *The History of Torture in England.* Montclair, N.J.: Patterson Smith, 1975.

Pepys, Samuel. *The Shorter Pepys: Selected and Edited by Robert Latham.* Berkeley: University of California Press, 1985.

Phelps Brown, E. H., and Sheila V. Hopkins. "Seven Centuries of the Prices of Consumables Compared with Builders' Wage-Rates." *Economica* 23, no. 92, new series (November 1956), pp. 296–314.

Porter, Roy. *English Society in the Eighteenth Century.* London: Penguin Books, 1990.

———. *London: A Social History.* Cambridge, Mass.: Harvard University Press, 1995.

Richardson, John. *The Annals of London.* London: Cassell, 2000.

———. *London and Its People.* London: Barrie & Jenkins, 1995.

Ruding, Rogers. *Annals of the Coinage of Britain and its Dependencies,* 3rd edition. Three volumes. London: John Hearne, 1840.

Saw, Reginald. *The Bank of England, 1694–1944.* London: George G. Harrap, 1944.

Schaffer, Simon. "Golden Means: Assay Instruments and the Geography of Precision in the Guinea Trade." In *Instruments, Travel, and Science: Itineraries of Precision from the Seventeenth to the Twentieth Century.* Edited by Marie-Noëlle Bourguet, Christian Licoppe, and H. Otto Sibum. London: Routledge, 2002.

———. "Newton on the Beach: The Information Order of *Principia Mathematica.*" Lecture delivered in several locations, text, private communication with the author, 2007.

———. "Newtonian Angels." Draft book chapter, private communication with the author, 2008.

Scheurer, P. B., and G. Debrock, eds. *Newton's Scientific and Philosophical Legacy.* Dordrecht, Netherlands: Kluwer Academic Publishers, 1988.

Schubert, H. R. *History of the British Iron and Steel Industry from c. 450 B.C. to A.D. 1775.* London: Routledge and Kegan Paul, 1957.

Schwoerer, Lois G., ed. *The Revolution of 1688–1689.* Cambridge: Cambridge University Press, 1992.

Shapin, Steven. *A Social History of Truth.* Chicago: University of Chicago Press, 1994.

———, and Simon Schaffer. *Leviathan and the Air-Pump.* Princeton, N.J.: Princeton University Press, 1985.

Shirras, G. Findlay, and J. H. (Sir John) Craig. "Sir Isaac Newton and the Currency." *Economic Journal* 55, no. 218/219 (June–September 1945), pp. 217–41.

Shoemaker, Robert B. *Prosecution and Punishment.* Cambridge: Cambridge University Press, 1991.

Spence, Joseph. *Anecdotes, Observations, and Characters, of Books and Men.* Edited by Samuel S. Singer. London, 1820.

Sprat, Thomas. *History of the Royal Society.* Edited with critical apparatus by Jackson I. Cope and Harold Whitmore Jones. St. Louis: Washington University Studies, 1958.

Stone, Lawrence, ed. *An Imperial State at War.* London: Routledge, 1994.

Sullivan, Robert E. *John Toland and the Deist Controversy.* Cambridge, Mass.: Harvard University Press, 1982.

Thayer, H. S., ed. *Newton's Philosophy of Nature.* New York: Hafner Publishing, 1953.

Tobias, J. J. *Crime and Police in England, 1700–1900.* Dublin: Gill and Macmillan, 1979.

Wales, Tim. "Lovell, Sir Salathiel," in the *Oxford Dictionary of National Biography.* Oxford: Oxford University Press, 2004.

Waller, Maureen. *1700: Scenes from London Life.* London: Hodder and Stoughton, 2000.

Wennerlind, Carl. "The Death Penalty as Monetary Policy: The Practice and Punishment of Monetary Crime, 1690–1830." *History of Political Economy* 36 no. 1 (2004), pp. 131–61.

———. Isaac Newton's Index Chemicus." *Ambix* 22 (1975), pp. 174–85.

———. "Newton's Marvelous Years of Discovery and Their Aftermath: Myth versus Manuscript." *Isis* 71, no. 1 (March 1980), pp. 109–21.

Whiteside, D. T. "The Prehistory of the *Principia.*" *Notes and Records of the Royal Society of London* 45, no. 1 (January 1991), pp. 11–61.

Whitfield, Peter. *London: A Life in Maps.* London: British Library, 2006.

Woolf, Harry, ed. *The Analytic Spirit.* Ithaca, N.Y.: Cornell University Press, 1981.

Index